D1248887

PROCESSING, STRUCTURE AND PROPERTIES
OF BLOCK COPOLYMERS

PROCESSING, STRUCTURE AND PROPERTIES OF BLOCK COPOLYMERS

Edited by

M. J. FOLKES

Department of Materials Technology,
Brunel University, UK

ELSEVIER APPLIED SCIENCE PUBLISHERS
LONDON and NEW YORK

ELSEVIER APPLIED SCIENCE PUBLISHERS LTD
Crown House, Linton Road, Barking, Essex IG11 8JU, England

Sole Distributor in the USA and Canada
ELSEVIER SCIENCE PUBLISHING CO., INC.
52 Vanderbilt Avenue, New York, NY 10017, USA

British Library Cataloguing in Publication Data

Processing, structure and properties of block
copolymers.
1. Block copolymers
I. Folkes, M. J.
547.8′4   QD382.B5

ISBN 0-85334-323-3

WITH 11 TABLES AND 85 ILLUSTRATIONS

Photoset in Malta by Interprint Limited
Printed in Great Britain by Galliard (Printers) Ltd, Great Yarmouth

Preface

Block copolymers represent an important class of multi-phase material, which have received very widespread attention, particularly since their successful commercial development in the mid-1960s. Much of the interest in these polymers has arisen because of their rather remarkable microphase morphology and, hence, they have been the subject of extensive microstructural examination. In many respects, the quest for a comprehensive interpretation of their structure, both theoretically and experimentally, has not been generally matched by a corresponding enthusiasm for developing structure/property relationships in the context of their commercial application. Indeed, it has been left largely to the industrial companies involved in the development and utilization of these materials to fulfil this latter role. While it is generally disappointing that a much greater synergism does not exist between science and technology, it is especially sad in the case of block copolymers. Thus these materials offer an almost unique opportunity for the application of fundamental structural and property data to the interpretation of the properties of generally processed artefacts.

Accordingly, in this book, the editor has drawn together an eminent group of research workers, with the specific intention of highlighting some of those aspects of the science and technology of block copolymers that are potentially important if further advances are to be made either in material formulation or utilization. For example, special consideration is given to the relationship between the flow properties of block copolymers and their microstructure. This relates closely to the development of mechanical anisotropy during processing, which is discussed in the context of both block copolymers and their blends with homopolymers. Considerable possibilities exist for the practical application of this anisotropy in the design of advanced rubber systems having specified properties. The link between the mechanical properties of idealized test specimens and practical moulded artefacts is achieved through the use of comparatively elementary fibre reinforcement theory. This treatment serves as a salutary reminder in showing how studies of model composites can be used to interpret the mechanical properties of more generally processed components. This is complemented by an extensive discussion of the segmented copolymers, with special emphasis on segmented polyurethanes, which clearly shows the important modifications to the properties of these block copolymers that can result from changes in their molecular architecture.

Above all, it is hoped that this book will prompt further thought and activities in this important area of polymer science and engineering and assist in the gradual evolution of a rational design basis for block copolymer systems.

M. J. FOLKES

Contents

List of Contributors

S. ABOUZAHR

Polymer Materials and Interfaces Laboratory, Department of Chemical Engineering, Virginia Polytechnic Institute and State University, Blacksburg, Virginia 24061, USA.

S. L. AGGARWAL

The General Tire & Rubber Company, Research Division, Akron, Ohio 44329, USA.

R. G. C. ARRIDGE

H. H. Wills Physics Laboratory, University of Bristol, Royal Fort, Tyndall Avenue, Bristol BS8 1TL, UK.

M. J. FOLKES

Department of Materials Technology, Brunel University, Kingston Lane, Uxbridge, Middlesex UB8 3PH, UK.

A. KELLER

H. H. Wills Physics Laboratory, University of Bristol, Royal Fort, Tyndall Avenue, Bristol BS8 1TL, UK.

J. LYNGAAE-JØRGENSEN

Instituttet for Kemiindustri, The Technical University of Denmark, Building 227, DK-2800 Lyngby, Denmark.

J. A. ODELL

H. H. Wills Physics Laboratory, University of Bristol, Royal Fort, Tyndall Avenue, Bristol BS8 1TL, UK.

G. L. WILKES

Polymer Materials and Interfaces Laboratory, Department of Chemical Engineering, Virginia Polytechnic Institute and State University, Blacksburg, Virginia 24061, USA.

CHAPTER 1

Introduction and Overview

S. L. AGGARWAL

The General Tire & Rubber Company, Akron, Ohio, USA

1. INTRODUCTION

The intent of this chapter is to provide a broad perspective of the chemistry and technology of block copolymers and their characteristic properties. It should serve as an overview of the structural features of block copolymers, and as an introduction to the subsequent chapters which treat in depth various topics concerning processing, structure and properties of block copolymers.

2. HISTORICAL

The rapid development of block copolymers from the early basic studies on organometallic compounds as polymerization catalysts, to the present level of commercial production, is astounding indeed. Extensive work from academic and industrial laboratories from all over the world has contributed to this development, and references to the technical literature are well documented in several reviews and recent books.[1-14] Readers of this chapter are urged to study particularly the reviews[9-14] in order to obtain a broad perspective of the chemistry and of the early studies on the structure of block copolymers.

The conceptual hypothesis of block copolymer structure may be traced to Alfrey *et al.* In their book on copolymerization[15] they pointed out that the presence of long sequences of a particular monomer in a copolymer may result in incompatability at a submicroscopic level. They predicted that such materials may show properties quite different from either a random copolymer or a macroscopic blend of the corresponding two homopolymers. The path of discoveries and developments from such a hypothetical prediction to the present state of the development of block copolymers as important materials, of both scientific and commercial interest, is a fascinating one. It has been highlighted in one of the recent publications.[16] Four milestones along this path of discoveries and developments stand out and are described below.

2.1. 'Living Polymers'

In 1956, Professor M. Szwarc and co-workers reported that certain anionic polymerization systems with sodium naphthalene resulted in 'living polymers' with anionically reactive ends to which a second monomer may be added without termination so that block copolymers may be prepared.[17-19] In the polymers that contain butadiene, the sodium naphthalene di-initiators, that were used by these workers in their studies, resulted in the 1:2 or 3:4 addition structures, which had some properties that were undesirable in commercial application. Thus, these studies did not attract sufficient interest in industrial laboratories.

2.2. Domain Structure

Bateman[20] and Merret[21] in 1957 observed that graft polymers of methyl methacrylate (MMA) on natural rubber (NR) may have two distinctly different physical forms, depending upon the solvent from which these polymers were precipitated. If precipitated from a solvent which was a 'good' solvent for poly(methyl methacrylate) and thus resulted in extended chains, but was a 'poor' solvent for NR resulting in collapsed NR chains, then the material was hard and non-tacky. However, if on the other hand the solvent was such that the reverse was true, i.e. extended NR segments but collapsed PMMA segments, the material was soft and flabby. Both physical forms were stable even under heavy milling. The vulcanizates prepared from these two forms had different properties, and they were different from those of the mixtures of the corresponding homopolymers.[20] Merret explained these differences by the postulate that the dry polymer consists of domains of the collapsed polymer chains as a distinct phase in a continuous matrix of the polymer that had existed in the extended chain form in the solution from which it was precipitated. At the time, it was an imaginative postulate for the two phase structure of the graft polymers in the solid state. It turned out to be a characteristic feature of the graft and block copolymers.

2.3. Alkyl Lithium Catalysts

The polymerization of isoprene with lithium metal as catalyst to prepare high *cis* polyisoprene was pioneered in the laboratories of The Firestone Tire & Rubber Company by Stavely *et al.*[22] It was extended to the use of alkyl lithium compounds as catalyst. Concurrent studies in several laboratories, including the Firestone laboratories,[23-29] led to the conclusion that alkyl lithiums are more suitable catalysts than lithium metal for the polymerization of such monomers as dienes and styrene. Further, these catalysts result in 'living polymerization' systems similar to that shown by Szwarc. The groups at The University of Akron under Professor Maurice Morton contributed significantly to the study of the anionic polymerization systems with alkyl lithium catalysts, and to the study of block copolymers.[30-35]

2.4. Thermoplastic Rubbers

Work from the Shell Laboratories in the USA resulted in the successful preparation of di- and triblock copolymers of butadiene and styrene and of isoprene and styrene, of well defined structures and compositions.[36-42] This work attracted worldwide interest, because one of the triblock copolymers of styrene and butadiene, called Kraton, containing about 70% butadiene and 30% styrene by weight had some interesting combinations of properties that had not been observed in polymer materials known until then.[42-45] It was a strong rubber capable of large reversible deformation, but required no external reinforcing fillers and vulcanization (crosslinking), as is the case for commonly vulcanized rubbers. Kraton could be processed as a thermoplastic, but had elastomeric properties at room temperature. Such block copolymers were thus called 'thermoplastic rubbers'. Historically, it is interesting to note that Kraton thermoplastic block copolymers resulted not so much from a concerted research to prepare block copolymers, but from a chance discovery from the work directed to reduce long time relaxation and flow (cold flow) in polybutadienes and polyisoprene prepared by using alkyl lithium catalysts, that was being studied in various laboratories and at Shell.[16] Adding styrene terminal blocks at each end of polybutadiene (or polyisoprene) molecules was tried as a solution to reduce cold flow in these polymers.

Tables 1 and 2 give the comparative strength and fatigue-to-failure properties, respectively, of Kraton thermoplastic rubber and typical vulcanized rubbers at room temperature.[46]

The workers from the Shell Laboratories also proposed a model to explain the strength and thermoplastic rubber properties observed with Kraton.[42] It is shown schematically in Fig. 1.[45,46] Electron microscopy

TABLE 1
COMPARATIVE STRENGTH VALUES[a] OF KRATON 101 THERMOPLASTIC
RUBBER AND TYPICAL VULCANIZED RUBBERS

	Tensile strength $(lb\ in^{-2})$	Elongation (%)
Triblock thermoplastic rubber (Kraton 101)	40 300	740
Natural rubber (carbon black reinforced and cured)	28 200	560
SBR synthetic rubber (carbon black reinforced and cured)	29 800	590

[a] Data from typical rubber formulations and at room temperature. Tensile strength values based on true cross-section at break $(1\ lb\ in^{-2} = 6.894 \times 10^3\ N\ m^{-2})$.

TABLE 2
COMPARATIVE FATIGUE-TO-FAILURE VALUES[a] AT APPROXIMATELY EQUAL
STRAIN ENERGY FOR KRATON THERMOPLASTIC RUBBER AND TYPICAL
VULCANIZED RUBBERS[46]

	Cycles to failure
Triblock thermoplastic rubber (Kraton 101)	145 000
Natural Rubber (carbon black reinforced and cured)	48 000
SBR synthetic rubber (carbon black reinforced and cured)	45 000

[a] Data with typical rubber formulations and at room temperature, using a Monsanto Fatigue Tester, ~110% elongation.

studies on microtomed thin sections or thin films cast from a solution of Kraton provided strong evidence for the general features of this model. Fortunately, an unrelated discovery of the use of osmium tetraoxide which stains the rubber phase was made at about this time, and it was of great value in studying the multi-phase structure of Kraton and similar block copolymers.[47] Using this staining technique, which makes rubber the electron dense phase (dark phase in the electron micrographs), the studies of a number of workers showed that in its essential features the proposed model was the correct one.[48-58] Figure 2 is one of the early electron micrographs from the author's laboratory which shows the

POLYSTYRENE   **POLYBUTADIENE**   **POLYSTYRENE**

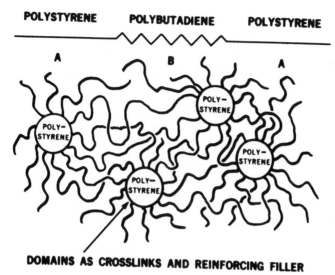

DOMAINS AS CROSSLINKS AND REINFORCING FILLER

FIG. 1.   Schematic of the domain structure of S–B–S triblock thermoplastic rubbers.[45,46]

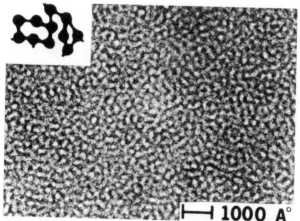

FIG. 2.   Electron micrograph of Kraton 101 cast from THF/MEK 90/10 by volume solution.[46,52]

spherical domains of ~ 100 Å dispersed in the polybutadiene matrix.[46,52] These studies laid the foundation for further extensive studies on block copolymers, and other multi-phase polymer systems.

3. CHAIN STRUCTURE AND BLOCK COPOLYMER TYPES

Block copolymers of controlled chain structure and composition are best prepared from the copolymerization of such monomers as styrene ($C_6H_5-CH=CH_2$) and butadiene ($CH_2=CH-CH=CH_2$), using anionic catalysts such as butyl lithium ($\overset{-}{Bu}.\overset{+}{Li}$). The two features of such polymerization systems that make it possible to prepare block copolymers of controlled structure are:

1. long life reactive ends, i.e. 'living polymers' as discussed above; and
2. a broad range of reactivity ratios for the copolymerization of these monomers, either in different solvents (e.g. benzene vs. tetrahydrofuran) or in the presence of different concentrations of weakly basic ethers.[59-64]

Figure 3 shows schematically the various block copolymer structure types that can be prepared.[46] In this figure, styrene units are shown as A (●) and the butadiene units are shown as B (○). The diblock copolymers may be prepared, using butyl lithium as catalyst, by first polymerizing styrene (A) to completion, and then polymerizing a sequence of butadiene (B) units on to the reactive end. The triblock ABA block copolymer may be prepared by adding monomer A, prior to termination, to the diblock copolymer with the active end; or the diblock copolymer reactive species may be coupled by a chemical coupling agent[65,66] to

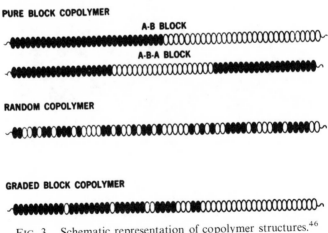

FIG. 3.   Schematic representation of copolymer structures.[46]

give an ABA triblock copolymer with a B block in the centre and an A block at each end. An especially convenient, and perhaps better controlled method, for the preparation of such triblock copolymers is by the use of the difunctional catalysts, such as the dilithio catalysts of such structure as:[67-69]

$$\overset{+}{\text{Li}} \quad \overset{-}{\text{CH}}-\text{CH}_2-\text{CH}_2-\overset{-}{\text{CH}} \quad \overset{+}{\text{Li}}$$
$$\underset{\text{C}_6\text{H}_5}{|} \qquad\qquad \underset{\text{C}_6\text{H}_5}{|}$$

The central block is formed by the polymerization of monomer B, and then the blocks at each terminal end are added on by the polymerization of monomer A.

The graded block structures (or tapered blocks, as they are sometimes referred to) result when the comonomers are added together initially, and the reactivity ratios are controlled by carrying out the polymerization in benzene containing a suitable concentration of a weakly basic ether.[59-64,70]

The concept of block copolymers is not limited to diblock, triblock and graded block copolymers. Multi-block copolymers containing several sequences of A and B chemical units are prepared either by design or as a result of the chemistry involved in their synthesis. The block copolymers of styrene and butadiene are of special interest, because of the structural control that is possible, and the well-defined chain structures that can be obtained and analysed. They serve as models for studying structure–property relationships of block copolymers. Thus in the following discussion on the characteristic properties of block copolymers, reference is made mostly to styrene/butadiene block copolymers.

4. CHARACTERISTIC PROPERTIES OF BLOCK COPOLYMERS

4.1. Microphase Separation and Morphology of Domains

Microphase separation in the solid state of block copolymers comprised of imcompatible blocks is one of the most characteristic properties of block copolymers. Figure 2 is a typical electron micrograph of a triblock copolymer of styrene/butadiene containing about 30% styrene and 70% butadiene by weight. Even though it is not evident in most electron micrographs of block copolymers, both thermodynamic considerations and increasing experimental evidence show that the boundaries between the domains cannot be sharp.[70-78] There is an interfacial region in which mixing of phases occurs. The morphology and nature of the separated domains, including the diffuseness and size of the phase boundaries depend on chemical composition and chain structure, molecular weights of the constituent blocks, thermodynamic interaction parameters, casting solvent and solidification conditions, and temperature. Theoretical and quantitative aspects of microphase separation in block copolymers are discussed in Chapter 3 of this book. The range of solid state structure, i.e. the nature of the morphology of the dispersed and continuous phase, is well illustrated by the following examples for the block copolymers of styrene and butadiene, or isoprene in place of butadiene.[46,79-81]

4.1.1. Effect of Chemical Composition
The electron micrographs in Fig. 4 show the effect of the increasing weight percentage of styrene on the morphology of the segregated phases. In the block copolymer that has butadiene as the major component, the dispersed phase is that of interconnected small spheres or short rods. As the amount of styrene increases, the morphology changes to that of a lamellar structure of polystyrene and polybutadiene layers. Finally, as the amount of styrene in the block copolymer increases further, and polystyrene becomes the dominant component, the dispersed phase is that of polybutadiene in a matrix of polystyrene.

4.1.2. Effect of Chain Structure (Pure vs. Graded Block)
The electron micrographs in Fig. 5 show the differences in morphology between the 'pure' and 'graded' triblock copolymers of 75% styrene and 25% isoprene by weight. The diffuse boundary of the dispersed phase in the graded block copolymer results from the favourable conditions for phase mixing in block copolymers of such structure.

4.1.3. Effect of Solvent Used in Preparation and Isolation
The two electron micrographs in Fig. 6 show the difference in the morphology of a triblock ABA polymer of 40% styrene (A) and 60% isoprene (B): one prepared in toluene that is a good solvent for the polystyrene terminal blocks; and the other prepared in heptane which is a good solvent for the central block and a poor solvent for the terminal blocks. In the former, the polystyrene domains of short rods or interconnected spheres are of large dimensions and are not as well segregated as in the case of the polymer prepared in and isolated from heptane.

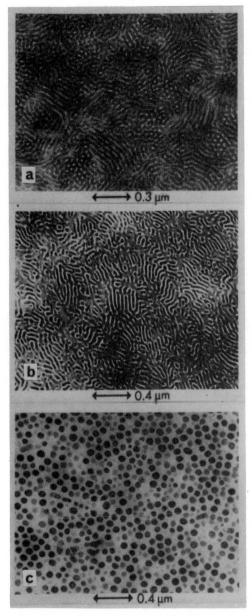

FIG. 4. Electron micrographs showing the effect of chemical composition on the domain structure of S–B–S block copolymers.[46] (a) 25 Styrene/75 butadiene; (b) 60 styrene/40 butadiene; (c) 90 styrene/10 butadiene.

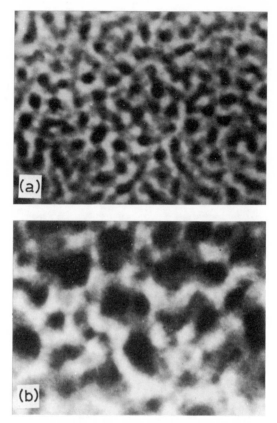

FIG. 5.   Electron micrographs showing the morphology of (a) 'pure' and (b) 'graded' S–I–S block copolymers.[46]

4.1.4. Effect of Molecular Architecture (Triblock vs. 'Star Block' Copolymers

Block copolymers in which, instead of linear architecture of the ABA triblock copolymers, there are several block segments radiating from a centre, have also recently been prepared, notably by Fetters and his associates.[82–84] Architecture of such 'star block' copolymers has a marked effect on the regularity and morphology of the segregated phase. The electron micrograph in Fig. 7 shows the morphology of the styrene domains in a film from a sample of styrene/isoprene radial block copolymer containing 30% styrene, an average number of 15 arms with

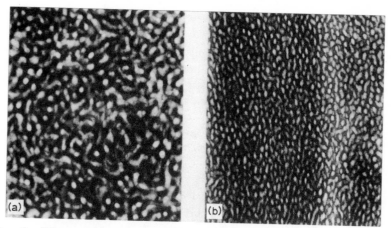

FIG. 6. Effect of the solvent used for polymerization and isolation on the morphology of a S–B–S block copolymer: (a) toluene; (b) heptane.[46]

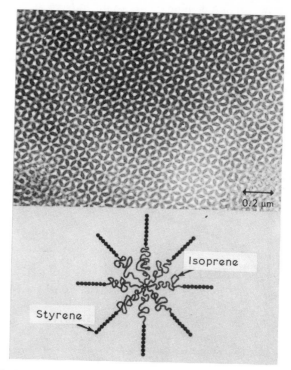

FIG. 7. (Top) Electron micrograph showing the morphology of polystyrene domains in a film of a 'radial' block copolymer of styrene and isoprene. (Bottom) Schematic representation of the architecture of such a polymer.[46]

each arm having a molecular weight of 71 000.[46] At the bottom of this figure is shown schematically the architecture of such a polymer. The noteworthy point is the characteristic regularity of the size and arrangement of the domains, even in a film cast from solution.

The morphology of the block copolymers and the effect of the processing conditions on morphology are discussed in detail in Chapter 2 by Professor Keller and Dr Odell.

5. PLASTIC-TO-RUBBER TRANSITION, AND DUCTILITY OF GLASSY DOMAINS

In general, elastomeric block copolymers, which have the two phase (or multi-phase) solid state structure with glassy domains in a matrix of an elastomer, show strain softening, i.e. a strain induced plastic-to-rubber transition. This behaviour has been shown by a number of studies[46,48,52,64,85-89] as characteristic of such block copolymers. Similar behaviour is also observed for the blends of block copolymers with the corresponding plastic homopolymer of molecular weight lower than, or comparable to, the molecular weight of the block copolymer component.

Figure 8 is a stress–strain plot of a film of the triblock styrene/but-

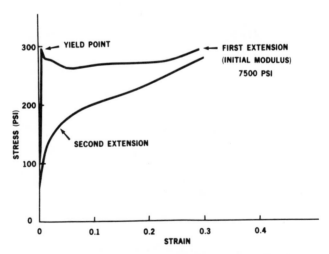

FIG. 8.   Stress–strain curve of Kraton 101 S–B–S triblock copolymer cast from THF/MEK (90/10 by volume) solution.[46,100]

adiene block copolymer used to obtain the electron micrograph shown in Fig. 2. Closer examination of the electron micrograph in Fig. 2 shows that the polystyrene domains are loosely interconnected to form a swivel-like structure, shown schematically on the left-hand corner of Fig. 2. The stress–strain curve in Fig. 8 shows that during the first elongation cycle, the material has a high modulus, typical of a plastic rather than that of a rubber. There is a yield point at about 3% elongation, beyond which the stress remains essentially constant with further elongation up to almost 200% strain. This typical plastic-like behaviour is characteristic of such systems. When the applied stress exceeds the yield stress, necking, i.e. cold drawing, appears at a localized region in the specimen, and this grows continuously until the whole specimen is covered. Upon further stretching, the stress rises rapidly and fracture soon follows (this part is not shown in the stress–strain curve).

During the second and subsequent stress–strain cycles, the behaviour is typical of a crosslinked rubber. Interestingly, the plastic-to-rubber transition is reversible. After the strain has been removed the specimen 'heals' itself, and the original stress–strain behaviour returns.[52] The 'healing' is accelerated at elevated temperature, but is observed, though at a slower rate, at temperatures well below the glass transition temperature of the glassy polymer domains.

The plastic-to-rubber transition in these block copolymers has been

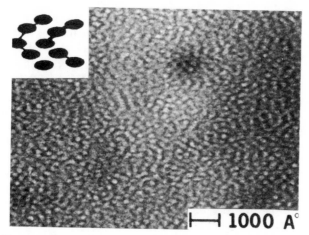

H—H 1000 A°

FIG. 9. Electron micrograph showing the deformation of polystyrene domains in S–B–S block copolymers on stretching 300%.[46,100]

interpreted as a consequence of the fragmentation of the lamellar or interconnected glassy domains on stretching. This breaking of the short-range order in the glassy domains is followed at high elongation by the ductile deformation of the domains, even at temperatures well below the glass transition temperature. Figure 9 is an electron micrograph of the sample used for the electron micrograph shown in Fig. 2, on stretching to about 300%. The spherical domains in the original sample are deformed to elliptical domains. The structural re-formation to the original structure (and thus to the original stress–strain behaviour) has been interpreted[88] as a consequence of: (1) the orientation of the elastomer segments (giving rise to decreased entropy) and (2) an increase in the interfacial energy resulting from the fragmentation of the glassy domains and breakdown of the short-range order. These two factors thus contribute to the interfacial domain boundary relaxations activated by the fragmentation of the glassy domains.

6. PROPERTIES AND MORPHOLOGY OF BLENDS OF BLOCK COPOLYMERS WITH HOMOPOLYMERS

Blends of block copolymers with homopolymers, corresponding to either or both of the blocks, provide a new dimension in the control of morphology of multi-phase structure. In some cases, such blends have resulted in materials with useful combinations of properties. Helfand[73,90–92] and Meier[93] have treated the statistical thermodynamics of such blends, and have made predictions of domain sizes and preferred morphology. Professor Kawai and his co-workers at the University of Kyoto have shown from their studies over several years that a homo-polymer added to a block copolymer may either be solubilized in the corresponding domains of the block copolymer, or it may act as an 'alloying agent' or surfactant, and thus contribute to the stabilization of a discrete homopolymer/block copolymer boundary.[53,56,94–99] In the for-mer case, the homopolymer imbibed is usually the minor constituent of the blend, and is of molecular weight less than that of the corresponding block of the copolymer. For the latter case, the homopolymer is the major constituent and has molecular weight either comparable to or higher than that of the corresponding block (in the ABA triblock copolymer it applies to that of terminal blocks A). Examples of this type of behaviour in blends of styrene/butadiene/styrene triblock copolymers with polystyrene homopolymer are shown in Figs 10 and 11.[100]

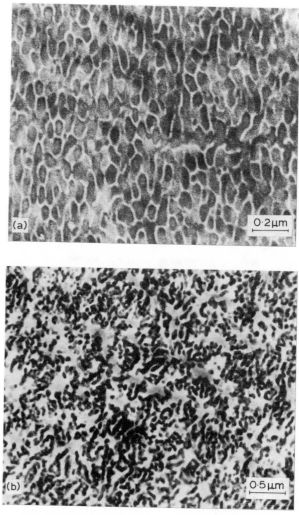

FIG. 10. Electron micrographs of; (a) a 60/40 styrene/butadiene triblock co-polymer; (b) a 50/50 blend of the block copolymer with homopolymer poly-styrene.[100]

In Fig. 10 are shown the electron micrographs of (a) a triblock copolymer containing 60% styrene and 40% butadiene, and (b) a 50/50 blend of this block copolymer with polystyrene homopolymer. In the block copolymer, the polystyrene domains are thin distorted lamellae

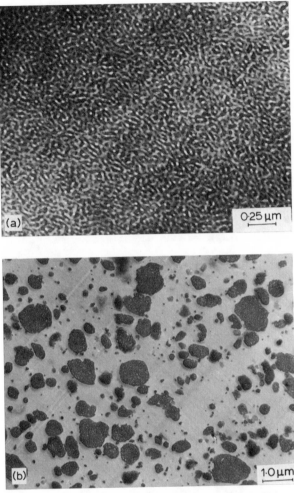

Fig. 11. Electron micrographs of: (a) a 40/60 styrene/butadiene triblock co-polymer; (b) a 75/25 blend of polystyrene homopolymer and the block copolymer.[100]

with a polybutadiene phase between them. In the blend, polystyrene becomes a continuous phase. The styrene blocks fuse into the polystyrene phase and act as an 'alloying agent'. There is no block copolymer with its original morphology in the blend. The butadiene block is now in the form of distorted cylindrical domains. Since the styrene blocks and homopolymer polystyrene are now a single phase, the polybutadiene domains would be well bonded to the matrix.

Figure 11(a) is an electron micrograph of an S–B–S triblock co-polymer containing 40% styrene and 60% butadiene. The composition of this block copolymer is the reverse of that discussed above. Since for comparable total molecular weights of the two block copolymers, the molecular weight of the terminal blocks is proportional to the com-position, the molecular weight of the terminal styrene blocks in this block copolymer is lower than that discussed above. The polystyrene domains in this case are in the form of short cylinders which show frequent interconnections.

Figure 11(b) is the electron micrograph of a blend of 25 parts of this block copolymer and 75 parts of polystyrene homopolymer. The disper-sed phase in this case is the block copolymer which has retained its original morphology. The styrene blocks of the block copolymer at the interface act, however, as an 'alloying agent' and provide good bonding between the phases. In such a blend, the block copolymer acts as a toughening agent for glassy polystyrene homopolymer.[101] It has greater impact strength than usual high impact polystyrene (HIPS) at compar-able levels of flexural modulus (Table 3).

TABLE 3

PROPERTIES OF BLENDS OF S–B–S TRIBLOCK COPOLYMERS WITH PS, HIPS, AND HOMOPOLYSTYRENE

Sample	Diene rubber in the system (%)	Notched impact strength, (ft-lb in^{-1})	Flexural modulus $\times 10^{-5}$ (lb in^{-2})	Heat distortion temperature (°C)
Polystyrene-1,	0	0·2	4·9	92
Polystyrene-2	0	0·25	4·7	101
HIPS-1	3–5	1·0	3·46	75
HIPS-2	8–12	1·7	2·1	79
75/25 PS/S–B–S triblock blend^a	15·0	7·5	3·0	87

^{a}Block copolymer, 40/60 S/B contents; $[\eta] = 3·01$ dl g^{-1}.

A good understanding of the mechanism of rubber toughening of glassy polymers such as polystyrene has evolved during the last 15 years. The recent book by Bucknall[103] and a review paper by Haward and Bucknall[104] summarize well the mechanism of toughening of glassy polymers by inclusion of an elastomer or elastomer-rich phase. In contrast to this, the work on the impact modification of glassy polymers, and the understanding of the mechanism responsible for toughening by block copolymers, is in its infancy.

6.1. Morphology of Toughened Glassy Polymers

Figure 12 is an electron micrograph of a high impact polystyrene sample (HIPS). Comparison of Fig. 12 with the electron micrograph in Fig. 11(b) for the 75/25 blend of polystyrene/block copolymer shows several similarities and several differences between the morphologies of these two multi-phase systems.

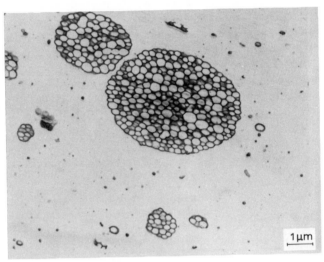

FIG. 12.   Electron micrograph showing the multi-phase structure of a high impact polystyrene (HIPS).[100]

In both, the dispersed phase is elastomer-rich; and in both the dispersed phase itself is a two-phase system with characteristic morphology. In both systems, an adequate bonding of the elastomer-rich phase to the surrounding matrix is achieved — a necessary but not sufficient condition for obtaining impact enhancement. In HIPS, interfacial adhesion is attained probably by the incorporation of polystyrene grafts on rubber molecules into the polystyrene matrix, and in the blends of block copolymers by the incorporation of some of the terminal blocks and phase mixing at interfacial boundaries.

The size of the dispersed phase in the block copolymer blend is of the order of few thousand ångströms, while the size of the dispersed phase in HIPS is up to ten thousand ångströms in diameter.

The polystyrene domain in the dispersed block copolymer phase of the blend is of submicroscopic dimensions, in contrast to the size of the dispersed phase in HIPS, in which large polystyrene domains encased in rubber are present.

Even though, at present, there is only a preliminary understanding of the underlying phenomena that result in the superior combination of high impact strength and high flexural modulus in the case of the blend of block copolymer and polystyrene, as compared with HIPS, some of the general observations may be useful in guiding future work on the development of high impact materials by blending with block copolymers.

6.2. Mechanism of Toughening

Two basic mechanisms contribute to the toughening of polystyrene by inclusion of an elastomer rich phase: (1) plastic deformation of the continuous polystyrene matrix phase, as a consequence of the hydrostatic pressure on the continuous matrix phase due to differences between its Poisson's ratio and that of the dispersed elastomeric phase; (2) stress concentration at points near the equator of the elastomeric phase with respect to the principal stress direction, which results in development of microcrazes. The dispersed phase also acts as a craze arrester. As deformation increases, it is important that the large or interconnected crazes do not grow throughout the system, and thus result in premature failure. Both mechanisms may contribute to the toughening of HIPS and the blends of the block copolymer and polystyrene.

The differences between the impact properties of the two materials are due to differences in the details of: (1) the deformation behaviour of the dispersed phase; (2) the optimum size of the dispersed phase required for efficient formation of microcrazes; and (3) the morphology of the microcrazes formed. Some of these differences in the multi-phase structure of the two materials — morphology of the microcrazes and deformation of the dispersed phases — are shown in the electron micrographs in Figs 13 and 14, of stretched thin films cast from toluene solution. The three electron micrographs in Fig. 13 are for HIPS; those in Fig. 14 are for the

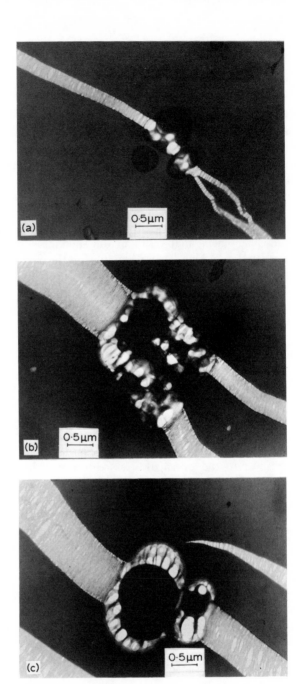

FIG. 13. Electron micrographs showing the crazes in a stretched film of high impact polystyrene (HIPS) Styron 495.[102]

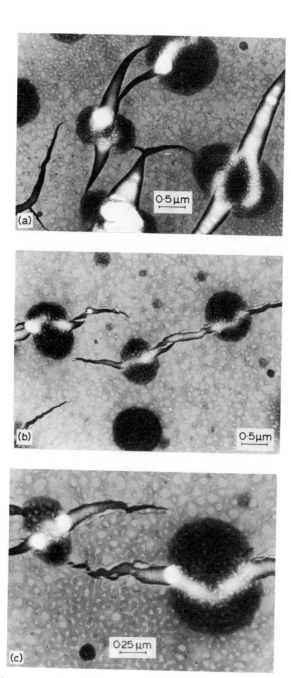

FIG. 14. Electron micrographs showing the crazes in a stretched film of the blend of block copolymer and polystyrene homopolymer.[102]

block copolymer blend with polystyrene.[94,102] The following are some of the noteworthy features:

1. Crazes are prominent in the electron micrographs of both the materials. In both, crazes initiate at the equator of the dispersed phase and in each case the dispersed phase undergoes considerable deformation. However, in the case of the block copolymer blend, there is considerable local yielding of the block copolymer particles. The yielding in this case is inhomogeneous; the thin areas, near the centre, become thinner without affecting the regions in the vicinity. This deformation is akin to the plastic-to-rubber transition discussed earlier as a characteristic property of the elastomeric triblock copolymers. The consequences of this are not only the dissipation of some energy by such deformation, but also the resultant stress concentration near the equator of the particles, thus initiating microcrazes. The block copolymer phase thus acts as an efficient microcraze initiator in these blends. In the case of HIPS, the dispersed elastomer phase undergoes almost uniform distortion with no local yielding or thinning.

2. In HIPS, some of the particles of the elastomer rich phase are quite large — up to ten thousand ångströms in diameter, as compared with the few thousand ångströms required for the most efficient rate of developing microcrazes.[94] Further the stress concentration is highest at areas with large radius of curvature. The microcrazes in HIPS thus tend to be initiated at the truncated points of the irregularly shaped dispersed phase. There is a tendency for the crazes to form a wide band of crazed material. For the block copolymer blend, this does not seem to be the case. These microcrazes have a characteristic shape, wider and highly extended near the phase boundary, and narrower and less extended away from it.

6.3. Use of Block Copolymers as Blending Agents

Elastomeric block copolymers as blending agents for improving the impact resistance of glassy, or in some cases, crystalline polymers are attracting considerable interest, and have heretofore unrealized potential. Composition and the molecular weight of the block copolymers used for making such blends is very important in order to obtain materials with controlled morphology and solid state structure, and thus a useful combination of functional properties.

TABLE 4

NOTEWORTHY CLASSES OF BLOCK POLYMERS OF CHEMICAL COMPOSITION DIFFERENT FROM THOSE OF STYRENE AND BUTADIENE

Block A	Block B	References
Poly(tetramethylene terephthalate)	Poly(tetramethylene ether)	105–107
Polysulphone	Polydimethylsiloxane	108,109
Polycarbonate	Poly(dimethylene siloxane)	110
Polyethylene	Polypropylene	111
Soft polyurethane segment[a]	Hard polyurethane segment[a]	112–115

[a] The soft segment structure shown is for that prepared from the reaction of MDI (diphenylmethane-4,4'-diisocyanate) with poly(propylene ether glycol) oligomer. The hard segment is for that prepared from the reaction of MDI with ethylene glycol. Combination with other isocyanates, glycols, and polyether glycols provides segmented polyurethanes of a wide range of structural variation.

7. SCOPE OF AVAILABLE BLOCK COPOLYMERS

The block copolymers of styrene and dienes are the most investigated model block copolymer materials. They sparked considerable interest in block copolymers of different chemical composition and molecular architecture (star-shaped, comb-shaped, and others beyond the linear chain structures discussed above). The recent book by Noshay and McGrath[7] gives a good survey of the block copolymers of different chemical structure and composition. The main classes of block copolymers that have been developed and are of current technological interest are listed in Table 4. The blocks identified as A and B may be combined in various combinations including multi-block, $(AB)_n$, structures to optimize combinations of properties desired for their useful applications.

REFERENCES

Books for Supplementary Reading
1. Molau, G. E. (Ed.), (1971). *Colloidal and Morphological Behavior of Block and Graft Copolymers*, Plenum Press, New York.
2. Aggarwal, S. L. (Ed.), (1979). *Block Polymers*, Plenum Press, New York.
3. Sperling, L. H. (Ed.), (1974). *Recent Advances in Polymer Blends, Grafts, and Blocks*, Plenum Press, New York.
4. Allport, D. C. and Janes, W. M. (Eds), (1973). *Block Copolymers*, Applied Science Publishers Ltd, London.
5. Ceresa, R. J. (Ed.), (1972). *Block and Graft Copolymerization*, Wiley, New York.
6. Burke, J. J. and Weiss, V. (Eds), (1973). *Block and Graft Copolymers*, Syracuse University Press, New York.
7. Noshay, A. and McGrath, J. E. (1977). *Block Polymers*, Academic Press, New York.
8. Goodman, I. (Ed.), (1982). *Developments in Block Copolymers — 1*, Applied Science Publishers Ltd, London.

Selected Review Articles for Supplementary Reading
9. Dreyfuss, P., Fetters, L. J. and Hansen, D. R. (1980). Elastomeric block polymers, *Rubber Chem. Technnol.*, **53**(3), 728.
10. Estes, G. M. and Cooper, S. L. (1970). Block polymers and related heterophase elastomers, *J. Macromol. Sci. Reviews Macromolecular Chem.*, **C4**(2), 313–36.
11. Morton, M. and Fetters, L. J. (1977). Anionic polymerizations and block polymers, in: *High Polymers Series*, Volume 29, (Ed. C. E. Schildknecht and I. Skeist), Wiley Interscience, New York.

12. Aggarwal, S. L. (1976). Structure and properties of block polymers and multiphase polymer systems: An overview of present status and future potentials, *Polymer*, **17**, 938–56.
13. Hashimoto, T., Shibayama, M., Fujimura, M. and Kawai, H. (1981). Microphase separation of block polymers, *Mem. Fac. Eng., Kyoto Univ.*, **43**(2), 184–223.
14. Shen, M. (1979). Properties and morphology of amorphous hydrocarbon block copolymers, *Adv. Chem. Ser.*, **176** 181–204.

Specific Literature References

15. Alfrey, T., Bohrer, J. J. and Mark, H. (1952). *Copolymerization*, Interscience, New York.
16. Legge, N. R., Holden, G., Davison, S. and DeLaMare, H. E. (1975). In: *Applied Polymer Science*, (Ed. J. Kenneth Craver and R. W. Tess), Organic Coatings and Plastics Chemistry Division of the American Chemical Society, Washington, DC.
17. Szwarc, M. (1956). *Nature*, **178**, 1168.
18. Szwarc, M., Levy, M. and Milkovich, R. (1956). *J. Am. Chem. Soc.*, **78**, 2656.
19. Szwarc, M. (1968). *Carbanions, Living Polymers, and Electron Transfer Processes*, Wiley, New York.
20. Bateman, L. C. (1957). *Ind. Eng. Chem.*, **49**, 704.
21. Merret, F. M. (1957). *J. Polym. Sci.*, **24**, 467.
22. Stavely, F. W. *et al.* (1956). *Ind. Eng. Chem.*, **48**, 778.
23. Foreman, L. E. (1969). *Polymer Chemistry of Synthetic Elastomers*, Part II, (Ed. J. P. Kennedy and E. G. M. Tornquist), Wiley, New York.
24. Adams, H. E., Bebb, R. L., Forman, L. E. and Wakefield, L. B. (1972). *Rubber Chem. Technol.*, **45**(5), 1252.
25. Crouch, W. W. and Short, J. N. (1961). *Rubber Plastics Age*, **42**, 276.
26. Railsback, H. E., Beared, C. C. and Haws, J. R. (1964). *Rubber Age*, **94**, 583.
27. Hsieh, H. L. and Glaze, W. H. (1970). *Rubber Chem. Technol.*, **43**(1), 22.
28. Zelinski, R. P. and Childers, C. W. (1968). *Rubber Reviews*, **41**, 161.
29. Szwarc, M. (1973). In: *Block and Graft Copolymers*, (Ed. J. J. Burke and V. Weiss), Syracuse University Press, New York, pp. 1–16.
30. Morton, M., McGrath, J. E. and Juliano, P. C. (1969). *J. Polym. Sci.*, **C26**, 99.
31. Morton, M. (1970). In: *Block Polymers*, (Ed. S. L. Aggarwal), Plenum Press, New York, pp. 1–10.
32. Morton M. (1971). In: *Encyl. Polym. Sci. Technol.* Vol. 15, (Ed. H. F. Mark), Interscience, New York, Supplement, p. 508.
33. Fetters, L. J. (1972). *J. Elastoplast.*, **4**, 34.
34. Fetters, L. J. (1973). In: *Block and Graft Copolymerization*, Vol. 1, (Ed. R. J. Ceresa) Wiley, New York, p. 99.
35. Morton, M. and Fetters, L. J. (1975). *Rubber Chem. Technol.*, **48**(3), 359–406.
36. Milkovich, R., South African Patent 280,712 (1963). Assigned to Shell Oil Company.
37. Porter, L. M., US Patent 3,149,182 (1964), filed October, 1957.

38. Bailey, J. T., Bishop, E. T., Hendricks, W. R., Holden, G. and Legge, N. R. (1965). 'Thermoplastic elastomers, physical properties and applications', paper presented at a meeting of the Division of Rubber Chemistry, American Chemical Society, Philadelphia. Work summarized in *Rubber Age*, (Oct. 1966), 69–74.

39. Holden, G. and Milkovich, R., US Patent 3,265,765 (1966). Assigned to Shell Oil Company.

40. Bailey, J. T., Bishop, E. T., Hendricks, W. R., Holden, G. and Legge, N. R. (1966). *Rubber Age*, **98**, 69.

41. Holden, G., Bishop, E. T. and Legge, N. R. (1967). In: *The Proceedings of the International Rubber Conference*, Maclaren, London.

42. Holden, G., Bishop, E. T. and Legge, N. R. (1969). *J. Polym. Sci.*, **C26**, 37.

43. Smith, T. L. (1970). In: *Block Polymers*, (Ed. S. L. Aggarwal), Plenum Press, New York, pp. 137–52.

44. Smith, T. L. and Dickie, R. A. (1969). *J. Polym. Sci.*, **C26**, 163.

45. Aggarwal, S. L. (1979). *Shell Polymers*, **3**(2), 43.

46. Aggarwal, S. L. (1976). *Polymer*, **17**, 938.

47. Kato, K. (1965). *J. Electronmicroscopy (Japan)*, **14**, 220; (1967). *Polym. Eng. Sci.*, **7**, 38.

48. Hendus, H., Illers, K. M. and Ropte, E. (1967). *Kolloid-Z.*, **216/217**, 110.

49. Matsuo, M., Ueno, T., Horino, H., Chujyo, S. and Asai, H. (1968). *Polymer*, **9**, 425.

50. Henderson, J. F., Grundy, K. H. and Fisher, E. (1968). *J. Polym. Sci.*, **C16**, 3121.

51. Wilkes, Garth L., and Stein, R. S. (1969). *J. Polym. Sci.*, (*A2*), **7**, 1525.

52. Beecher, J. F., Marker, L., Bradford, R. D. and Aggarwal, S. L. (1969). *J. Polymer Sci.*, **C26**, 117.

53. Inoue, T., Soen, T., Hashimoto, T. and Kawai, H. (1970). In: *Block Polymers*, (Ed. S. L. Aggarwal), Plenum Press, New York, pp. 53–79.

54. Lewis, P. R. and Price, C. (1971). *Polymer*, **12**, 258.

55. Inoue, T., Moritani, M., Hashimoto, T. and Kawai, H. (1971). *Macromolecules*, **4**, 500.

56. Uchida, T., Soen, T., Inoue, T. and Kawai, H. (1972). *J. Polym. Sci.* (*A2*), **10**, 101.

57. McIntyre, D. and Campos-Lopez, E. (1970). In *Block Polymers*, (Ed. S. L. Aggarwal), Plenum Press, New York, pp. 19–30.

58. LeFlair, R. T. (1971). *XXIIIrd International Congress of Pure and Applied Chemistry*, Vol. 8, Butterworth, London, p. 195.

59. Kelly, D. J. and Tobolsky, A. V. (1959). *J. Am. Chem. Soc.*, **81**, 1597.

60. Tobolsky, A. V. and Rogers, C. E. (1959). *J. Polym. Sci.*, **40**, 73.

61. Spirin, Yu. L., Polyakov, D. K., Gantmakber, A. R. and Medvedev, S. S. (1962). *Polym. Sci. USSR*, **3**, 233.

62. Korotkov, A. A. and Rakova, G. V. (1962). *Polym. Sci. USSR*, **3**, 990.

63. Livigni, R. A., Marker, L. F., Shkapenko, G. and Aggarwal, S. L. (1967). Structure and transition behavior of isoprene–styrene copolymers of different sequence length. In: *Struct. and Properties of Elastomers*, Am. Chem. Soc., Division of Rubber Chemistry Symposium, Montreal, Canada.

64. Aggarwal, S. L., Livigni, R. A., Marker, L. F. and Dudek, T. J. (1973). In:

Block and Graft Copolymers, (Ed. J. J. Burke and V. Weiss), Syracuse University Press, New York, pp. 157–94.
65. Zelinski, R. P. and Childers, C. W. (1968). *Rubber Reviews*, **41**, 161.
66. Adams, H. E., Bebb, R. L., Forman, L. E. and Wakefield, L. B. (1972). *Rubber Chem. Technol.*, **45**, 1252.
67. Reed, P. J. and Urwin, J. R. (1972). *Organomet. Chem.*, **39**, 1.
68. *Analysis, handling, and reactions of difunctional organolithium initiators*, Product Bulletin 194, 'Dili,' Lithium Corporation of America, Bessemer City, USA.
69. Fetters, L. J. and Morton, M. (1969). *Macromolecules*, **2**, 169.
70. Yasuhisa, T., Nakamura, N., Hashimoto, T. and Kawai, H. (1980). *Polym. J. (Japan)*, **12**(7), 455.
71. Meier, D. J. (1969). *J. Polym. Sci.*, **C26**, 81.
72. Meier, D. J. (1973). In: *Block and Graft Copolymers*, (Ed. J. J. Burke and V. Weiss), Syracuse University Press, New York, pp. 105–20.
73. Helfand, E. (1975). *Macromolecules*, **8**, 552.
74. Helfand, E. and Wasserman, Z. R. (1978). *Macromolecules*, **11**, 960.
75. Hashimoto, T., Nagatoshi, K., Todo, A., Hasegawa, H. and Kawai, H. (1974). *Macromolecules*, **7**, 364.
76. Hashimoto, A., Todo, A., Itoi, H. and Kawai, H. (1977). *Macromolecules*, **10**, 377.
77. Hashimoto, T., Shibayama, M. and Kawai, H. (1980). *Macromolecules*, **13**, 1237.
78. Tsukahara, Y., Nakamura, N., Hashimoto, T. and Kawai, H. (1980) *Polym. J. (Japan)*, **12**(7) 455.
79. Matsuo, M., Ueno, T., Horino, H., Chujyo, S. and Asai, H. (1968). *Polymer*, **9**, 425.
80. Uchida, T., Soen, T., Inoue, T. and Kawai, H. (1972). *J. Polym. Sci. (A2)*, **10**, 101.
81. Matsuo, M. (1968). *Japan Plast.*, 6.
82. Bi, Le-Khac, Fetters, L. J. and Morton, M. (1974). *Polymer Preprints: Am. Chem. Soc., Div. Polymer Chem.*, **15**(2), 157.
83. Bi, Le-Khac and Fetters, L. J. (1975). *Macromolecules*, **8**, 98.
84. Bi, Le-Khac (1975). Ph.D. Dissertation, University of Akron, Akron, USA.
85. Henderson, J. F., Grund, K. F. and Fisher, E. (1968). *J. Polym. Sci., Polym. Symp.*, **16** 3121.
86. Fischer, E., Henderson, J. F. (1969). *J. Polym. Sci., Polym. Symp.*, **26**, 149.
87. Hong, S. D., Shen, M., Russell, T. and Stein, R. S. (1977). In: *Polymer Alloys*, (Ed. D. Klempner and K. C. Frisch), Plenum Press, New York, pp. 77–95.
88. Hashimoto, T., Fujimura, M., Saijo, K., Kawai, H., Diamant, J. and Shen, M. (1979). *Adv. Chem. Ser.*, **176**, 257–75.
89. Fischer, E. (1968). *J. Macromol. Sci., Chem.*, **A2**, 1285.
90. Helfand, E. (1974). In: *Recent Advances in Polymer Blends, Grafts and Blocks*, (Ed. L. H. Sperling), Plenum Press, New York, pp. 141–56.
91. Helfand, E. (1975). *Accounts Chem. Research*, **8**, 295.
92. Helfand, E. (1975). *J. Chem. Phys.*, **62**, 999.

93. Meier, D. J. (1974). *Appl. Polym. Symp.*, **24**, 67; (1974). *Am. Chem. Soc.*, *Polym. Prepr.*, **15**, 171.
94. Kawai, H., Hashimoto, T., Miyoshi, K., Uno, H., and Fujimura, M. (1980). *J. Macromol. Sci., Phys.*, **B17**(3), 427–72.
95. Kawai, H., Soen, T., Hashimoto, T., Ono, T. and Uchida, T. (1971). *Mem. Fac. Eng., Kyoto Univ.*, **33**, 383.
96. Kawai, H. and Inoue, T. (1970). *Japan Plast.*, (July), 12–20.
97. Inoue, T., Soen, T., Kawai, H., Fukatsu, M. and Kurata, M. (1968). *J. Polym. Sci.*, **B6**, 75.
98. Inoue, T., Soen, T., Hashimoto, T. and Kawai, H. (1970). *Macromolecules*, **3**, 87.
99. Inoue, T. (1971). Colloidal and morphological behavior of block copolymer and its bulk properties, Dissertation to Department of Polymer Chemistry, Faculty of Engineering, Kyoto University, Kyoto, Japan.
100. Aggarwal, S. L. and Livigni, R. L. (1977). *Polym. Eng. Sci.*, **17**, 498; also reprinted in (1978). *Rubber Chem. Technol.*, **51**(4), 775.
101. Durst, R. R., Griffith, R. M., Urbanic, A. J. and van Essen, W. J. (1976). *Adv. Chem. Ser.*, **154**, 239–46.
102. Aggarwal, S. L. (1977). US–Japan Joint Symposium on Elastomers, Univ. of Akron, Ohio, October 17–21, including some unpublished electron micrographs made by Horst Oswald of The General Tire & Rubber Company, Research Division, Akron, Ohio, USA.
103. Bucknall, C. B. (1977). *Toughening Plastics*, Applied Science Publishers, London.
104. Haward, R. N. and Bucknall, C. B. (1976). *Pure Appl. Chem.*, **46**, 227.
105. Knox, J. B. (1973). *Polym. Age*, **4**, 357.
106. Brown, M. and Witsiepe, W. K. (1972). *Rubber Age*, **104**, 35.
107. Witsiepe, W. K. (1973). In: *Polymerization Reactions and New Polymers*, (Ed. N. A. J. Platzer), American Chemical Society, Washington, DC, p. 39.
108. Noshay, A., Metzner, M. and Merriam, C. N. (1971). *Polym. Prepr.*, **12**, 1.
109. Robeson, L. M., Noshay, A., Metzner, M. and Merriam, C. N. (1973). *Agnew. Makromol. Chem.*, **29/30**, 47.
110. Kambour, R. P. (1970). In: *Block Polymers*, (Ed. S. L. Aggarwal), Plenum Press, New York, p. 263.
111. Kontos, E. G., Esterbrook, E. K. and Gilbert, R. D. (1962). *J. Polym. Sci.*, **61**, 69.
112. Saunders, J. H. and Frisch, K. C. (1962). *Polyurethanes: Chemistry and Technology*, Wiley Interscience, New York.
113. Trappe, G. (1968). In: *Advances in Polyurethane Technology*, (Ed. J. M. Buist and H. Gudgeon), Wiley Interscience, New York, Ch. 2 and 3.
114. Wright, P. and Cumming, A. P. C. (1969). *Solid Polyurethane Elastomers*, Maclaren, London, Ch. 2.
115. Puett, D. (1967). *J. Polym. Sci.*, **25**(A), 839.

CHAPTER 2

The Interrelation between Microstructure and Properties of Block Copolymers

A. KELLER and J. A. ODELL

H. H. Wills Physics Laboratory,
University of Bristol, UK

1. INTRODUCTION

The purpose of this article is to present a summarizing survey of work done at Bristol on a particular block copolymer system during the period 1971–1977. By its very nature such an article may appear restricted in scope in several respects. First, it is essentially restricted to material from one laboratory only; secondly, (as will be apparent) to basically one particular copolymer system, and to essentially one facet of this system, namely to the relation between microstructure and properties — and this at a time when there is much more recent activity in this topical field. In the face of all this we justify the present article on the following four counts:

1. The material comprised by it is a totally self-contained story which has never been told before in the framework of a single article. Even if there exist partial reviews dating from the time when the work was still in progress, these do not contain the full material.[1-3]
2. We consider the material important, particularly when viewed in its totality. The reason for this is that all the diverse constituents fit together harmoniously in remarkable agreement. This we feel is instructive in its own right as an example of practically complete consistency in structure/property relations, setting a model for this kind of study for composite materials in general and for multicomponent polymers, especially block copolymers.
3. The structures revealed, the properties displayed, and their interrelation, are highly interesting in their own right. To our knowledge no corresponding systems have been identified since with the same state of practically idealized perfection.

4. Most importantly and forward-looking, the material in question should represent a firm basis to build on for future work. Indeed current revival of this work, now of 5–12 years standing, is, in the hands of one of its original contributors, in the process of producing new and definite results which have explicit technological implications.[4] These last mentioned ongoing works feature in another chapter of this book for which the present article is thus preparing the ground.[5]

The polymer system in question is a three-block copolymer of polystyrene/polybutadiene/polystyrene, more explicitly defined in the formula below and to be referred to as S–B–S with an overall composition ratio of $S = 0.25$ and $B = 0.75$ (weight fraction) with molecular weights of 10^4 and 5.5×10^4, respectively.

$$[CH_2-CH=CH-CH_2]_n-[CH_2-CH]_m-[CH_2-CH=CH-CH_2]_n$$

Before embarking on the actual subject matter the following possibly well-known preliminaries will be recapitulated (e.g. Refs 6 and 7). The S and B constituents being incompatible, segregate into separate phases. As, however, the constituents within a given chain are molecularly connected, this phase segregation will be restricted to submicroscopic dimensions. In the particular case of S and B constituents, S is glassy and B is rubbery at room temperature, and thus we have a glass–rubber composite. When B is the majority component it will form a continuous rubbery matrix containing dispersed glassy particles of S. Owing to the molecular connectedness between the phases the glassy particles act as physical crosslinks within a rubber, thus forming a network, a state in which such materials find their most widespread application. Nevertheless, they differ significantly from conventional chemically crosslinked elastomers, in as far as the network-forming effect of the physical crosslinks is removable above T_g of the glassy phase, where the material becomes mouldable; hence the term 'thermoelastomers' for this family of materials.

The morphology of the microphases takes up several specific forms according to the volume ratios of the constituents. In particular, up to 12% volume fraction the minority component is in the form of uniform spheres, and between 12 and 35% as uniform cylinders of practically infinite length, in both cases the majority components forming the continuous matrix. In the composition range of 35–65% of either component the microphases are of lamellar character, or looking at it another way, when composition ratios are comparable the distinction between matrix and dispersed phase disappears. For the same constituent pairs (e.g. S and B in the present case) the role of matrix and particulate phase reverses beyond the 65% composition for the initially minority component. Thus, in the case of S–B–S we pass from a crosslinked rubber to a toughened polystyrene with the relative increase of the S component; the glassy S phase becomes the matrix for S content beyond 65%.*

The origin of the particular microphase structure in question has become the subject of intensive theoretical studies. The earlier ones have featured in some of our previous reviews.[2] There are more recent ones of great importance (e.g. Ref. 9). However, we shall not be further concerned with them here but proceed to the issue of microstructure and mechanical properties.

It had been observed before we entered the field that the submicroscopic microphase particles can group themselves into organized arrays over sizeable portions of an electron microscopic field of view. The starting point of our own involvement was the recognition that it was possible to obtain samples where this organization extended over the entire macroscopic specimen, over say a plug of 1 cm diameter and several centimetres length. Such a sample could therefore be considered as a macroscopic 'single-crystal'. These 'single-crystals' were very unconventional. Namely, on the atomic and molecular scale they were entirely amorphous, the repeating entity constituting the lattice being the amorphous microphase particles themselves with dimensions of the order of 10 nm. This curiosity apart, such samples offered a unique opportunity, both for characterizing the microphase structure and for studying the

---

*Although the sphere–cylinder–lamellar sequence is widely held to be valid we remark here that the regime of the spheres may include more complex but very specific high symmetry network structures. This is based on the little heeded works on lipids by Luzzati et al.[8] and invoked by ourselves in Bristol in recent work on S–B–S, where we were unable to explain our observations on the basis of the spherical microphase structure (see later in the section on swelling).[30]

influence of this structure on properties. This arises from the fact that any microstructurally significant direction or section could be identified in terms of a macroscopic direction or cross-section, and vice versa. thus creating a direct correspondence between microstructure and macroscopic behaviour. The rest of this article will be concerned with this correspondence. The significance of this topic, as we see it, has been enumerated above. To this we wish to add that in general the single-crystal thus identified represents the basic unit of more complicated textures such as arise in technological processing. In the knowledge of the behaviour of this basic unit and of the texture of a given article (i.e. the arrangement of such single-crystals within a given block copolymeric object) the properties of such an object can be predicted; or, alternatively, properties, as measured, can be interpreted in a rational and unequivocal manner. Further, and more forward-looking, such knowledge opens the way to the planned achievement of desirable properties by appropriate control of the fabrication variables, or alternatively enables the interpretation and eventual removal of undesirable features such as might arise. As already stated all these latter exploitations of the present works to be reviewed here, have already been put into practice as to be described in Chapter 4.

The principal sample morphology of our present concern will be that of an S–B–S copolymer of 25% S and 75% B (per weight) composition, giving rise to glassy S cylinders within a rubbery B matrix. Briefer mention will be given to lamellar composites arising from closely equal S:B ratios and a single reference to the 'inverse' structure from a copolymer of 70:30 S:B ratio.

2. HISTORICAL — IDENTIFICATION OF S–B–S 'SINGLE-CRYSTALS'

The Bristol involvement had its origin in the low angle X-ray examination of an extruded plug of S–B–S, Kraton K102, itself originating from the researches at Genoa University. This first examination, originally a spot test, was due to Dr Pedemonte, who while staying in Bristol in connection with other researches, availed himself of our low angle X-ray facilities to test a particular model of microphase segregation considered at the time. Surprisingly, instead of the diffuse scatter anticipated, discrete reflections in the form of spots were apparent at low angles, indicating the existence of regular Bragg spacings and beyond that, a

single-crystal texture.[10,11] Following up this initial chance observation
the definition of the reflections was improved; first by heat-treatment of
the plugs, and subsequently by fabricating new plugs with the specific
aim of perfecting the structure. Here, first the X-ray evidence as obtained
on our well-developed single-crystal type structure will be presented, and
in the section to follow a brief indication of the method established for
achieving optimum regularity within the samples.

2.1. The X-ray Evidence for Single-crystals

The essential evidence can be summed up by Figs 1(a) and (b). In Fig.
1(a) the X-ray beam was along the plug axis, hence extrusion direction, in
Fig. 1(b) perpendicular to it.

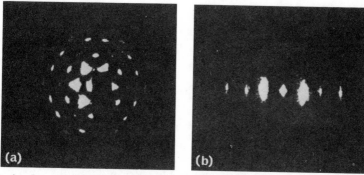

FIG. 1. Low angle X-ray diffraction patterns from a single-crystal sample of
extruded and annealed S–B–S copolymer (Kraton 102): (a) beam parallel to the
extrusion direction (plug axis); (b) beam perpendicular to the plug axis, which is
vertical.[10,11]

It is evident that the pattern in Fig. 1(a) represents a hexagonal lattice,
seen in the plug axis direction, of considerable regularity as evidenced by
the visibility of up to 4 orders (3 in the figure) of the hexagonal net in
reciprocal space. The same regularity is apparent in Fig. 1(b), although
confined to a single line of reflection perpendicular to the plug axis. It
will be immediately evident even from this one pair that we are dealing
with a parallel array of cylinders packed in a regular hexagonal lattice.
As our resolution limit was beyond 100 nm it follows from Fig. 1(b) that
the cylinders must be much longer than that, and for our purposes can
be considered as infinite, and for the same reason without any structure

along their length. The single-crystal character is apparent over at least the area comprised by the beam ($\sim 0.01\,\mathrm{mm^2}$) and further work has shown that in suitably perfected preparation this 'single-crystal character' can comprise practically the full macroscopic specimen, which as stated above can be up to 1 cm in diameter (with some region of lattice disorder remaining along the plug centre). The hexagonal cell parameter, i.e. centre-to-centre separation of the cylinders, is 30 nm. It follows from the stoichiometry and from the macroscopic densities of the individual phases that the cylinders are 15 nm in diameter. The same conclusion is reached from the intensities of the reflections, in particular from the systematic weakening of the 2nd order of the basic hexagonal reflexions. This means that the basic cylinder separation is 15 nm, i.e. the cylinders are separated by matrix material of a thickness which is about equal to their own diameter.

2.2. Fabrication of Single-crystal Samples

It was obvious from the outset that extrusion was responsible for the single-crystal nature of the sample. In subsequent works conditions for obtaining samples of maximum regularity and size were optimized.[12] This comprised an initial 'melting' at 100–120°C, then squeezing the viscous material through an orifice into a tube of cylindrical or rectangular cross-section (the latter being more suitable for some of the mechanical work to follow). This tube was heated along the walls to temperatures between 50–80°C so as to enable the material to flow slowly within it, allowing new material to enter into the tube. After plugs with cylindrical or rectangular cross-section of the desired length were obtained, the samples were cooled to ambient temperature and removed from the surrounding tube. Examination by low angle X-ray scattering revealed that the optimum regularity in the single crystal character was achieved along the peripheral annulus with a less regular central zone. By suitable optimization of the heating and flow conditions the extent of the latter could be confined to a minimum central core. In all circumstances, subsequent heat annealing could greatly improve the perfection of microphase packing order. Apart from providing a practical recipe the following noteworthy physical factors emerged. The cylindrical microphases and the corresponding macro lattice itself were not produced by the method of fabrication, as this was present in a less developed and randomly oriented form in the initial raw material. The function of the above process was to align the cylinders and enable them to pack in crystallographic register over large volumes. The most effective process for alignment within the whole procedure was the slow shear flow along the walls, and for the achievement of crystallographic register the subsequent thermal conditioning applied.

3. ELECTRON MICROSCOPIC VERIFICATION OF THE SINGLE-CRYSTAL TEXTURE

The 'lattice' as revealed by the low angle X-ray pattern had a periodicity of 30 nm and was inferred to be constituted of 15 nm diameter cylinders. This means that both lattice periodicity and the actual structure element forming it are comfortably within the range of low and medium resolution electron microscopy. In addition, the single-crystal character on the scale of the macroscopic plug enabled the structure to be viewed along any preselected direction. This opportunity was exploited in the transmission electron microscopic examination which followed.[13]

Samples of required thickness were obtained by sectioning at low temperatures where the B phase had glassy character, the direction of sectioning having been chosen according to the direction of viewing required. The contrast was obtained by selectively staining the unsaturated B phase by osmium tetroxide. The cutting was performed along two directions: transversely and longitudinally with respect to the hexagonal (hence the assumed cylinder) axis. Figure 2 shows a detail of a transverse and Fig. 3 that of a longitudinal section, where the dark regions represent the stained B phase. The hexagonal arrangement of light, hence S circles in Fig. 2, is immediately apparent. This, in combination with the parallel lines in the longitudinal sections such as in Fig. 3, is fully consistent with the cylindrical structure and the hexagonal lattice array deduced from the X-ray photographs. Allowing for variations within the photographs (see also below) the dimensions of both lattice periodicity and cylinder diameter are in satisfactory quantitative agreement with deductions from the X-ray diffraction patterns.

Tilting experiments were also performed both on the transverse and longitudinal sections while in the electron microscope with results which are in full agreement with the postulated structure. Tilting of the transverse sections (Fig. 4) led to streaking of the light circles all along the same direction perpendicular to the tilt axis. This has its origin in the fact that the sections in these samples had appreciable thickness (~ 100 nm), i.e. well in excess of the cylinder diameter, and thus in the tilted position the cylinder portions traversing the section were seen in an

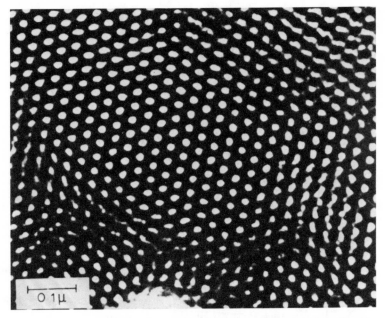

FIG. 2. Ultramicrotome section cut perpendicular to the extrusion direction of the S–B–S plug. Here as in subsequent photographs, the dark regions correspond to the stained B matrix.[13]

off-axial view. Even more interesting was the tilting of the longitudinal sections, such as that in Fig. 3, around the striation direction: here the striations became visible with high contrast only at 60° tilt intervals; in the intervening positions the striations were less distinct and hardly perceptible midway, i.e. at 30°, to positions of optimum contrast (Fig. 5). In fact, historically the viewing of any randomly positioned longitudinal section was first highly disappointing for lack of detail, until the requirement for tilting was recognized. The above requirement for tilting and the 60° tilt intervals involved, are fully consistent with the hexagonal lattice arrangement of cylinders within a section which is thick compared with the lattice periodicity. This follows directly from the expected imaging conditions of lattices, both from simple geometrical and from diffraction optical considerations, and in fact offers quite a unique example where both considerations can be applied. From the geometrical point of view, clearly we need to look along the direction of rows of cylinders, when we shall see both the lines of overlapping cylinders and the lines of the

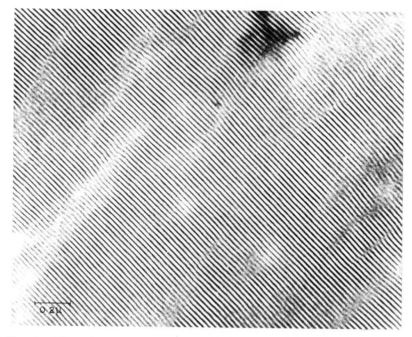

FIG. 3. Ultramicrotome section cut parallel to the extrusion direction of a single-crystal sample of extruded and annealed Kraton 102, showing striated structure.[13]

unobstructed gaps in between, in analogy with, say, a regular plantation of trees. Clearly such a view will be afforded at 60° intervals. Taking the diffraction optical view, in order to see a lattice spacing we need to recombine at least the first corresponding maximum in the diffraction pattern with the primary beam. In the hexagonal lattice within our longitudinal section the appropriate reflection will only appear when diffraction conditions are satisfied, which is at every 60° rotation around the hexagonal axis.

In addition to confirming the hexagonal lattice and its constituent cylindrical repeat units, certain faults are visible in Fig. 3. These are of two sources: (i) An occasional terminating lattice row forming an edge dislocation. These are rather rare considering the area of regions which are unfaulted in this manner. While interesting in themselves, these terminations do not invalidate the basic tenet that the cylinders can be considered as essentially of infinite length for our purposes. (ii) Some

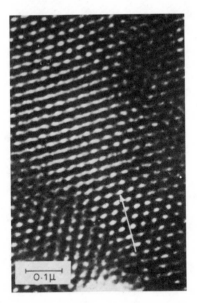

FIG. 4.   Portion of the same area as Fig. 2 (identifiable by the white area, part of a hole at bottom centre) but tilted by 12° around the arrow direction.[13]

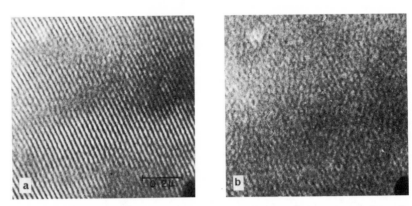

FIG. 5.   Ultramicrotome sections cut parallel to the extrusion direction of the S–B–S plug. (a) Shows an untilted section, while (b) shows a section tilted by 30° around the rod direction in which the contrast has disappeared; a further tilt of 30° restores the appearance of (a).[13]

fuzzy zones, extending transversely to the striations, where the striation contrast is impaired. These are regions of the sections where the above-defined viewing conditions are not exactly satisfied as could be verified by appropriately tilting around the striation direction when such regions can 'clear up' (naturally at the expense of others which become fuzzy instead). It is not clear to what extent such zones of altered lattice orientation (around the cylinder direction) are intrinsic to the sample or are due to distortions caused by sectioning: the latter is most likely a significant, if not the only, factor. An example of complete lattice perfection in transverse section, of gradually improved plug and section preparation technique, is illustrated by Fig. 6.

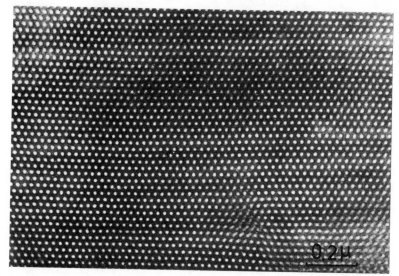

FIG. 6. Perpendicular section cut by the improved technique, involving pre-hardening and cooling and leading to the thinnest sections with optimum perfection.[14]

All the above aspects of longitudinal sections relate to the fact that the section thickness comprises at least a few repeat periods. While making the viewing of the lattice more definite, this feature deprives us from seeing individual cylinders as such, as these become visible only when overlapping. For certain issues, however, the side-on viewing of individual cylinders is required, as, for example, when the breakage of cylinders is being related to mechanically induced yielding (see later). To achieve this the section thickness had to be reduced. This was accomplished technically by combining Kato's method[15] of hardening the matrix through treatment with osmic acid and our cooling method applied previously. In this way longitudinal sections, sufficiently thin so as to comprise only one layer of cylinders, could be obtained.[14] The tilting experiment analogous to that in Fig. 4 now yields clearly defined striations for *all* rotation angles around the cylinder direction (Fig. 7), with the cylinder diameter (white zones) unaffected by the rotation but the rod centre-to-centre distance reduced in good quantitative agreement

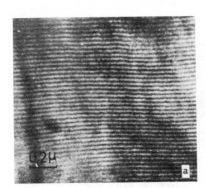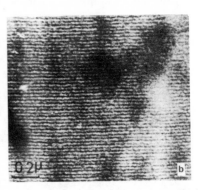

FIG. 7. Parallel sections prepared by method of Fig. 6. (a) Shows an untilted section, while (b) shows a section tilted by 30° around the rod direction (horizontal).[14]

with the foreshortening expected in the corresponding tilted position. Thus a stage has been reached where the structure of single cylinders can be meaningfully examined. Conversely, this also represents a concomitant test for assessment of sample thickness.

In summary, the single-crystal texture enabled realization of real space structural information about the block copolymer in question, in complete agreement with expectations from the X-ray diffraction pattern.

4. OPTICAL PROPERTIES

Under this heading we include birefringence and infra-red dichroism. Both of these properties are sensitive to the directionality of the appropriate material constituents, and hence reflect the characteristics of the microstructure. In the first place we checked how far the macroscopically determined optical (including spectroscopic) behaviour was consistent with the microstructure that has emerged from the preceding observations, and secondly, whether any additional information could be extracted from them.

4.1. Birefringence

Samples of single-crystal structure displayed optic uniaxial character with the plug axis as the optic axial direction, which is in accordance with expectations for a hexagonal lattice.[16] The birefringence was $(4.92 \pm 1.0) \times 10^{-4}$ with the largest refractive index along the plug axis (positive birefringence). Birefringence in polymers is normally associated with molecular orientation; hence potentially it should provide information about the molecular arrangement, as opposed to arrangement of the submicroscopic particles in our macro-crystal. Nevertheless, an oriented arrangement of dimensionally anisotropic particles, with at least one dimension smaller than the wavelength of light, can give rise to birefringence in itself without the involvement of molecular anisotropy, the so called 'form birefringence'. The present samples, consisting of uniformly oriented cylinders embedded in a matrix of differing refractive index, are expected to produce positive form birefringence. Consequently, before any conclusion can be drawn as regards molecular orientation the effect of form birefringence has to be assessed.

Form birefringence $n_a - n_0$ depends solely on the volume fraction (v_1, v_2) and refractive indices (n_1, n_2) of the two phases and can be expressed as

$$n_a - n_0 = \frac{v_1 v_2 (n_1^2 - n_2^2)^2}{2n_a [(v_1 + 1)n_2^2 + v_2 n_1^2]} \tag{1}$$

where

$$n_a^2 = v_1 n_1^2 + v_2 n_2^2$$

(see Ref. 16).

Here v_1 and v_2 are given and n_1 and n_2 are taken as the values measured on homopolymers. For our samples $n_a - n_0 = 5.15 \times 10^{-4}$ was calculated, in excellent agreement with the observed birefringence. In addition to the interesting fact that the usually elusive form birefringence has been quantitatively verified, this finding implies that there is no intrinsic molecular orientation in our single-crystals as prepared because the observed birefringence is accounted for by form birefringence alone (on deformation additional molecular birefringence is introduced, see later).

Form birefringence is expected to be altered in a predictable manner by introduction of selective swelling agents, which will affect the form birefringence by changing the refractive index difference and volume fractions. An example for this will be quoted in the section on swelling. The clear-cut information provided by the birefringence is obviously satisfying in itself. In addition, as will be seen in Chapter 4, the knowledge gained on the birefringence of a single-crystal provides a most effective vehicle for the understanding of texture and properties of more complex injection-moulded products, such as arise in technological fabrication. From the point of view of fundamentals it leads to the inference of an absence of measurable molecular orientation in either of the two phases. This is of basic importance for the subject of block copolymers, and, beyond that, has wider implications for polymer science. Namely, the absence of an overall orientation within an amorphous cylinder of 15 nm diameter makes it very difficult, if not impossible, to envisage larger scale structure elements within the glassy amorphous material — such as postulated as intrinsic constituents in amorphous polymers by some authors (e.g. Ref. 17) — as these would need to be randomized within the confines of the cylinder. Accordingly, our findings are against such structure models.

The importance of the issue of total molecular randomness calls for further independent evidence. Such is provided by infra-red dichroism, the results of which will be briefly quoted.

4.2. Infra-red Dichroism

In principle this method should be more specific than birefringence, as it should be possible to assess the orientation due to the S and B phases individually, by choosing suitable dichroic absorption bands characteristic of each constituent. Longitudinal sections of our single-crystal sample have shown no dichroism whatsoever for either of the two phases.[18] In fact the traces recorded with the polarizer parallel and perpendicular to the hexagonal axis (original plug axis) were coincident, which within our experimental sensitivity means that dichroism if present is less than 1·08.[18] It is difficult to define the limits on the molecular orientation set by this observation without a separate study of the infra-red dichroism of the homopolymers undertaken specially for this purpose. It is clear nevertheless that molecular orientation, if present, can only be very small in either of the phases. As already stated, this point is very relevant to theoretical considerations on the phase separation and size and shape of the dispersed particles.

As a caveat to the conclusion on the absence of any appreciable molecular orientation it needs emphasizing that this was deduced for the particular S–B–S single-crystal material with cylindrical microphase structure. It is not claimed that this conclusion is necessarily general for all microphase morphologies. In fact, as will be mentioned again in Section 9, small but assessable molecular orientations were deduced for block copolymers with lamellar microstructure, essentially along the same route via single-crystals as applied above for the cylindrical structures.

5. MECHANICAL BEHAVIOUR — DEFORMATION

Perhaps the most conspicuous consequences of the single-crystal nature of our samples are in the area of mechanical properties and deformation and in the interplay of the two. This arises from the fact that we have a composite system consisting of a rubbery constituent (compliant but elastic up to high levels of deformation) and a glassy constituent (stiff, but unable to support appreciable strain), and this in a geometrically uniquely specified combination, as established by the preceding structure research. It is this aspect of the material, presently under review, which is currently finding a practical outlet (see Chapter 4). In what follows this subject area will be presented in the following sequence. First, we confine ourselves to small strains within the elastic limits and consider the stress–strain relationship and modulus, to be followed by the study of the corresponding deformation by X-ray diffraction and then by birefringence. This will be followed by the effect of large strains involving yielding to comprise both stress–strain behaviour and the consequences for the structure.

6. MECHANICAL PROPERTIES IN THE ELASTIC RANGE (SMALL STRAINS)

The most conspicuous feature under this heading (and under those to follow) is the expectation of an extremely high degree of anisotropy, which should be displayed by a system consisting of a regular parallel array of glassy fibres embedded in a rubbery matrix of a modulus lower by several magnitudes, with the two components remaining in intimate molecular contact. It will be obvious even by qualitative considerations that we should expect the material to be very much stiffer in the fibre direction.

In order to examine this behaviour quantitatively,[16] test pieces were cut from a well-defined single-crystal specimen with the long dimension at different directions with respect to the plug axis, so as to enable the stress–strain behaviour to be recorded. The stress–strain curves and modulus E_θ were determined as a function of angle (θ) with respect to the original extrusion direction (Fig. 8). Figure 9 shows the results for some

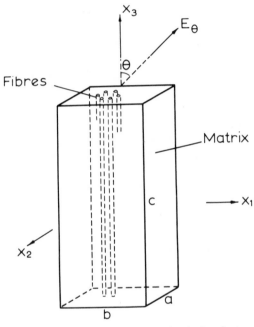

FIG. 8.   Definition of the axes describing mechanical anisotropy in transverse isotropic fibre reinforced materials, $E_{(\theta)} \equiv$ Young's modulus corresponding to direction defined by θ.

of the most significant angles, θ, while in Fig. 10 the corresponding E_θ values (together with some additional ones) are plotted. The very significantly larger stiffness along the cylinder direction $(\theta = 0°)$ as compared with that for larger θ values is immediately obvious. (Also, there is a small but definite minimum at $\theta = 55°$.) Thus, the sample behaves as a glass and as a rubber in two mutually perpendicular directions. This very large anisotropy is in agreement with the qualitative expectations laid

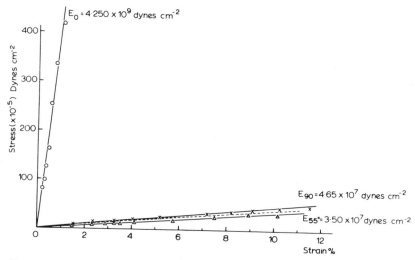

FIG. 9.   Stress–strain curves of single-crystal S–B–S copolymer (1 dyn cm$^{-2} \equiv$ 0·1 Pa). ○, stress parallel to extrusion direction; △, stress at 55° to extrusion direction; ×, stress perpendicular to extrusion direction. The broken line is a theoretical fit to E_{90}; for further details see Ref. 16.

out above. In what follows we shall compare actual numerical values with quantitative predictions.

To take the simplest model we consider the two phases as separate blocks without any interaction between them. Then the behaviour for $\theta = 0°$ will correspond to the case where the two blocks taken in the appropriate volume ratio are coupled parallel with respect to the stress. For this case the moduli will add, weighted with the appropriate volume fractions. Thus

$$E_0 = v_s E_s + (1 - v_s) E_b \qquad (2)$$

where v_s = volume fraction of polystyrene, E_s and E_b being the Young's moduli of S and B phases, respectively.

For E_s and E_b we take the macroscopically measured moduli of the corresponding homopolymer, which are 2×10^9 Pa and 1×10^6 Pa, respectively. (As $E_b \ll E_s$ the exact value for E_b, which depends on the degree of crosslinking in the polybutadiene homopolymer, is of little consequence.) The value calculated for E_0 is $4·0 \times 10^8$ Pa, in excellent agreement with the measured $4·25 \times 10^8$ Pa and fully supporting the underlying model.

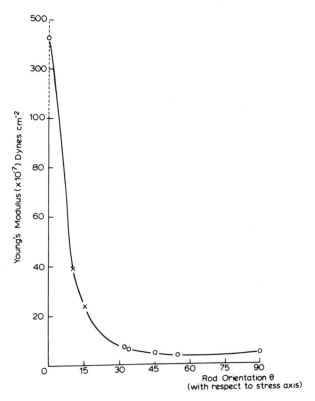

FIG. 10. Variation of Young's modulus, $E_{(\theta)}$, with orientation angle θ (1 dyn cm^{-2} ≡ 0·1 Pa). The curve is a least squares fit of eqn (6) to the experimental data: ○, from Ref. 16; ×, from Ref. 20.

Extension of the previous consideration to $\theta = 90°$ (i.e. stress perpendicular to the plug direction) would require a series coupling of the phases, i.e. here it is the compliances of the two phases, weighted according to their volume fractions, which should add. Bearing in mind that $E_b \ll E_s$ this gives

$$E'_{90} = \frac{E_b}{1 - v_s} \qquad (3)$$

However, it is apparent by simple considerations that here the independence of the two phases cannot be assumed, even as a first approximation, because the rubbery matrix stretched along $\theta = 90°$ will be prevented from contracting along the direction $\theta = 0°$ by the polystyrene cylinders. In fact the rubber will not be under uniaxial tension but under pure shear which will make it stiffer. The true modulus along $\theta = 90°$ will be given by

$$E_{90} = \frac{4}{3} E'_{90} = \frac{4}{3} \times \frac{E_b}{1 - v_s} \tag{4}$$

From the observed value of E_{90}, eqn (4) gives a value of $2·79 \times 10^6$ Pa for E_b, which is reasonable in view of the uncertainty of the degree of crosslinking to which our sample corresponds when compared with a standard homopolymer.

Significantly, a further check can be performed based on this simple model. It can be easily shown[16] that the restriction on the contraction of the rubber due to the cylinders will be removed when $\theta = 55°$; thus for this case eqn (3) should pertain:

$$E_{55} = \frac{E_b}{(1 - v_s)} = \frac{3}{4} E_{90} \tag{5}$$

Consequently, E_{55} should be lower than E_{90}; in fact it should be a minimum, which is immediately apparent from Figs 9 and 10. Further, the numerical relation between E_{55} and E_{90} can be tested. For $E_{90} = 4·65 \times 10^6$ Pa, a value for E_{55} of $3·48 \times 10^6$ Pa is expected from eqn (4), in practically complete agreement with the experimental value of $3·50 \times 10^6$ Pa.

In general, the angular dependence of the modulus of a uniaxially oriented sample should obey the tensor transformation formula (see Ref. 19):

$$\frac{1}{E_\theta} = S_{33} \cos^4 + (2S_{13} + S_{44}) \sin^2 \theta \cos^2 \theta + S_{11} \sin^4 \theta \tag{6}$$

where S_{ij} are the compliances (index 3 refers to the direction of axial symmetry, i.e. to $\theta = 0$, while 1 and 2 to any of the equivalent transverse directions (see Fig. 8)). With seven points of measurement and three unknown coefficients (S_{33}, ($2S_{13} + S_{44}$) and S_{11}) a least square test could be performed, and complete fit with experiment was obtained[16,20] (Fig. 10). Thus, internal consistency of the data is verified to a high level of accuracy.

The numerical values themselves can be accounted for by composite theory to which further reference will be made below, and which will be exhaustively treated in the article by Arridge and Folkes.[5] At this point we will simply state that even the simplest model theory provides a good approximation to the values determined experimentally.

We see that the basic 'crystal' unit of this type of material is mechanically extremely anisotropic. It will be apparent that these considerations will have to be taken into account when attempting to interpret the mechanical behaviour of a copolymer sample of a more complex 'polycrystalline' texture — as it may be more usual in practice — of which the present crystal is the basic texture unit.

7. INELASTIC DEFORMATION

As we have seen, for small strains S–B–S single-crystal texture material shows extreme mechanical anisotropy. Parallel to the cylinders the material is very stiff, whilst perpendicular to this direction rubbery behaviour is observed, the moduli in the two principal directions being in the ratio of about 100:1. These observations together with the full angular dependence of the modulus could be quantitatively accounted for by a simple composite model of the structure.[16]

In this section we examine the deformation behaviour in structural terms over an extended region of strain. There have been many works on the stress–strain behaviour of block copolymer systems.[21-23] A high degree of strain softening has been repeatedly observed, and in some cases, rehardening on subsequent sample treatments. From the outset such effects have been associated with the breakdown and re-formation of the microstructure.

The unique nature of the single-crystal material permits a new quantitative approach to these issues.[24] The deformation, associated stress–strain behaviour, and breakdown of the microstructure have been examined in two perpendicular directions as naturally afforded by the nature of the samples, namely perpendicular and parallel to the cylinder direction.

7.1. Strains Perpendicular to the Extrusion Direction

Low angle X-ray diffraction (LAXD) offers a ready technique for following structural changes on stressing a single-crystal texture sample perpendicular to the cylinder axes, as any deformation should be sensitively reflected by changes in the macrolattice and in the resulting diffraction pattern.

Figure 11 shows LAXD patterns corresponding to various strains

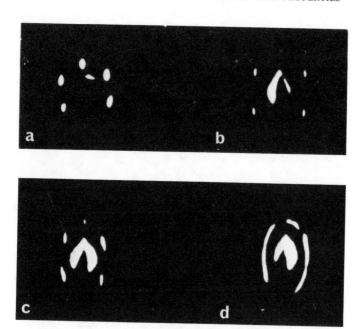

FIG. 11. LAXD pattern showing deformation of the macrolattice for strains applied perpendicular to the cylinder direction, with the X-ray beam parallel to the cylinders. Straining direction horizontal.[24] (a) 10%; (b) 15%; (c) 30%; (d) 50%.

applied perpendicular to the cylinder direction. The hexagonal array of reflections is progressively deformed — up to 20% strain. The lattice strain was found to be indistinguishable from the macroscopic sample strain, the inter-rod repeat spacing varying precisely in proportion to the sample dimensions (Fig. 12).

At greater strains the distorted hexagonal single-crystal LAXD pattern is progressively randomized, suggesting that the single crystal has broken up into randomly oriented fragments. The correspondence between lattice and sample strain is progressively lost; on increasing strains beyond 20%, a proportion of the applied strain is non-recoverable (Fig. 12). The sample strain which is in excess of lattice strain is associated with cracks which progressively develop in the material. This cracking as shown in Fig. 13 is a time-dependent process — even strains as low as 10% will result in the development of cracks if applied for several days. Remarkably, the cracks occur at 60° to each other in the sample, and correspondence with low angle X-ray diffraction shows that these are

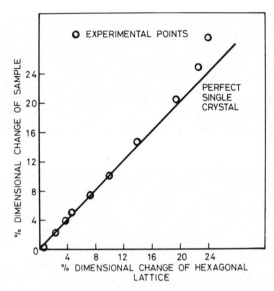

FIG. 12. The correspondence between deformation of the macrolattice and the applied strain perpendicular to plug axis.[24]

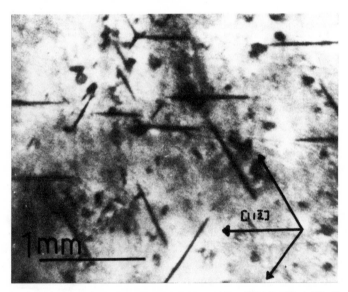

FIG. 13. Surface cracks in strained Kraton 102, with the $[1, 1, \bar{2}]$ direction superimposed[24] (strain perpendicular to plug axis).

well-defined crystallographic directions in the macrolattice ($[1,1,\bar{2}]$ directions in a two-dimensional Miller Bravais hexagonal lattice). The cracks in fact serve to identify the orientation of the macrolattice in an extruded plug.

The symmetry of the lattice shows that the $[1,1,\bar{2}]$ directions cross only highly strained links between polystyrene rods and so provide easy directions for crack propagation. As in the extruded material there are large areas of uncrosslinked rubber, those chains which do not directly link polystyrene cylinders cannot support the applied stress, except by chain entanglements. Thus some of the butadiene phase behaves as an uncrosslinked rubber, and flows slowly to comply with the imposed strain, explaining the time dependence of cracking. The unextruded non-single-crystal Kraton 102 has more dispersed polystyrene which is not in a single-crystal phase. As it does not have crystallographic directions along which cracks can easily propagate it does not develop cracks at low strain.

Measurements of the Young's modulus transverse to the cylinder direction agree well with those obtained from small strain experiments for strains of less than 10%; beyond this the observed modulus is progressively reduced, owing to the development of the cracks, as shown in Fig. 14.

In summary, owing to the single-crystal nature of the samples, deformations transverse to the cylinder axes could be isolated from deformations along other directions, and the source of reversible and irreversible effects identified in terms of the microstructure. In the last context the observed crack formation and its consequences are particularly illuminating in as far as they reveal the importance not only of the uniformly parallel alignment of the cylinders, but also the consequences of the fact that these cylinders are arranged on a regular lattice.

7.2. Large Strains Parallel to the Cylinders

There are special problems associated with the measurement of strain in samples of extreme mechanical anisotropy (the end-effect problem — see Chapter 4). A photographic method has been used to assess the distribution of surface strain and to provide a record of the yielding process.[24] There is some variability in the perfection of single-crystal extruded plugs, which particularly shows up in the inelastic behaviour. The structure determination methods mentioned previously were applied so as to select only the most perfect plugs for mechanical testing.

For strains less than 1·5%, the stress–strain curves show elastic

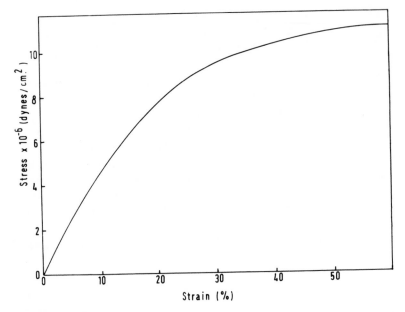

FIG. 14. Stress–strain curve for 0–60% strain perpendicular to the cylinder direction.[24]

behaviour, with no hysteresis, and with a Young's modulus close to that determined earlier in very small strain experiments. At larger strains (1·5–4%), the stress–strain curve becomes non-linear as shown in Fig. 15. On retesting after straining to 4%, the initial modulus is somewhat lower, a softening effect becoming increasingly apparent as the strain is increased (also reported by Pedemonte et al.[23]). Figure 16 shows a stress–strain curve for strains up to 110%. The material yields at about 3% strain, beyond which the load drops slightly. At higher strains the load remains constant as a neck propagates along the specimen, with a constant strain of 80% in the necked region and 1·2% in the unnecked region. When the sample has completely necked and the macroscopic sample strain has reached 80%, the load again increases and further deformation is elastic.

7.2.1. Recovery from Yielding
After complete necking upon removal of stress the material quickly relaxes to a residual sample strain of less than 0·5%. On immediate

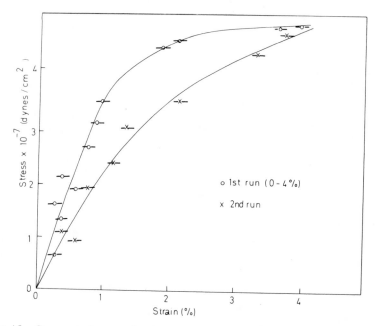

FIG. 15.   Stress–strain curve for 0–4% strain parallel to the cylinder direction.[24]

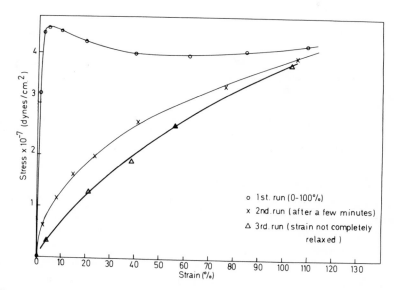

FIG. 16.   Stress–strain curves for 0–110% strain parallel to the cylinder direction.[24]

retesting the material shows rubber-like behaviour with a very low initial modulus (6×10^6 Pa).

After longer relaxation times, however, the initial modulus progressively recovers until eventually the whole pattern of yielding and neck propagation is re-established. Annealing above the polystyrene glass transition temperature in the relaxed state enables very rapid and almost complete recovery, the Young's modulus at low strains returning to 75% of its initial value. The effect is illustrated in Fig. 17. If, however,

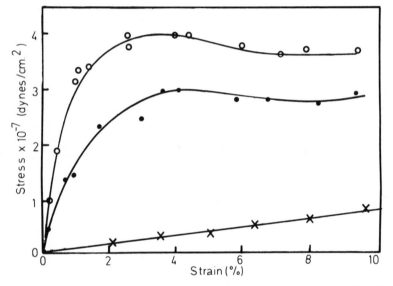

FIG. 17. Stress–strain curve for 0–10% strain parallel to the cylinder direction, showing the effect of annealing at 155°C after yielding. ○, initial test; ×, test after necking; ●, test after annealing of yielding material.[24]

the material is not allowed to relax completely but is held at some small strain ($\sim 1\%$), then there is no recovery, the material becomes permanently deformed and rubber-like behaviour persists indefinitely.

The implication of these results is that the microstructure of the material is disrupted by high strains, with a corresponding reduction in the Young's modulus along the extrusion direction. It seems that annealing above T_g for polystyrene enables the re-formation of the microstructure, restoring the mechanical properties. Remarkably, this process can also occur at room temperature, though much more slowly. These

structural inferences are in accord with suggestions from other observations of strain softening and rehardening.[23]

Thanks to the availability of the material in single-crystal form the
above implications can be verified in explicit terms by direct morphological examinations. Beyond that the single-crystals enable us to correlate
quantitatively structure, deformation, and properties, and further to
establish a model theory accounting for the experimentally established
correlations. The exposition of the relevant material will be the subject of
the ensuing sections.

7.3. The Morphology of Necked Material

7.3.1. Optical Properties

It has been established that virtually all of the observed birefringence in
single-crystal block copolymers in the undeformed state is due to form
birefringence,[16] and infra-red dichroism studies[18] support the contention
that there is little molecular orientation in either phase. However, since
the two phases have very different strain-optical coefficients, an examination of birefringence in the strained state can enable the distribution
of stress between the two components to be ascertained, with implications for the morphology of deformed material.

The birefringence has been measured as a function of strain for
elastically deformed material and through the neck region.[24] Any molecular orientation developed in the microphases will produce additional
birefringence. The orientation birefringence for different points in the
necked material can be estimated by subtracting the known form birefringence from the measured value. The form birefringence is found to be
quite small ($\sim 5 \times 10^{-4}$) compared with the strain-induced birefringence
($\sim 4 \times 10^{-3}$).

During elastic deformation the strain birefringence agrees well with
that expected for an affine deformation of the whole material (the lowest strain point on Fig. 18). Through the neck region, however, the
strain optical coefficient is progressively reduced until in completely necked
material it corresponds closely with that expected for deformation of the
polybutadiene phase alone. Overall the measured birefringence is entirely
consistent with the assertion that parallel to the rod direction the material
deforms uniformly until about 1·5% strain, beyond which the neck de-
velops.

The molecular orientation in the necked material is confined to the
polybutadiene phase; hence, the cylinders of polystyrene must be disrup-

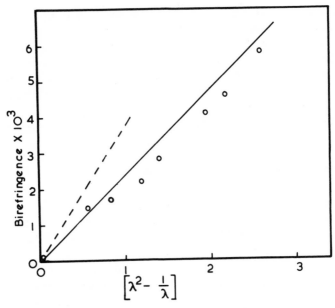

F IG. 18. Strain-induced birefringence plotted against $(\lambda^2 - 1/\lambda)$. -----, Predicted behaviour for affine deformation of the whole composite; ————, predicted behaviour for deformation of polybutadiene phase alone.[24]

ted. Qualitatively it is clear that broken cylinders bear less tensile load, which explains the observed stress softening (a more thorough analysis of the distribution of stresses will follow in Section 7.5).

7.3.2. Electron Microscopy of Strained Material

The precise nature of the microstructural changes during large deformations can be directly envisaged by the use of electron microscopy, a technique which has given definitive results in the course of previous works on unstrained material. However, certain additional difficulties present themselves. First, we know that necked material can recover its mechanical anisotropy on storage, suggesting that the continuity of the microphases can increase with time; this has been overcome by examining material which has been stored in the strained condition until the stress has relaxed, and thus a permanent deformation has occurred. Secondly, in order to see the discontinuities in the polystyrene cylinders, clearly electron microscope sections should be so thin that one looks through only one layer of cylinders. The techniques for achieving this,

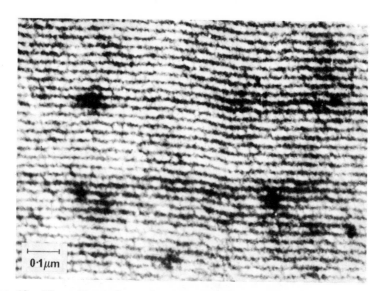

FIG. 19.   Ultra thin sections of necked material parallel to rods showing rod breakage.[14,24]

together with a test of thickness, have been described at the end of Section 3. The electron micrographs of the yielded samples clearly show breaks in the continuity of the polystyrene cylinders which are not present in the undeformed material. Figure 19 shows remaining rod lengths lying in the region of 70 to 110 nm. At higher magnification (Fig. 20) the presence of some very short rod lengths down to about 20 nm are also seen. Thus, because of the unique nature of these single-crystal samples, the quantitative disruption of the polystyrene phase can be verified by direct observation.

Before considering the quantitative interpretation of mechanical properties in terms of microstructure, an immediate observation on the recovery of modulus on storage may be made. Electron microscopy shows that the cylinders break up into short lengths of the order of 100 nm, but that these cylinders are still well aligned with each other. When the necked material is allowed to relax, the breaks close up and the strain is reduced to about 0·5%. For the observed cylinder lengths this corresponds to the broken ends of cylinders abutting with a gap of at most 0·5 nm. In view of the close proximity of broken ends, it is not surprising that even the very slow diffusion present in glassy polystyrene

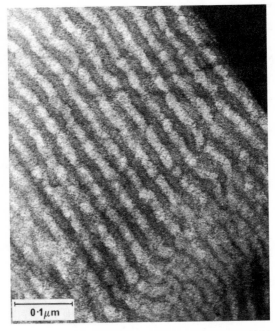

FIG. 20.   As Fig. 19 but at higher magnification.[14,24]

can reform the cylinders. For short relaxation times re-formation is incomplete, but can still result in a considerable improvement in the initial modulus; however, the stress required to rebreak the imperfect cylinders is very low, the consequent strain softening reoccurs at a very low stress. For longer relaxation times, or annealing above T_g for polystyrene, the cylinders were observed to re-form well and the material again displays yield and necking on repeated stressing. If the material has suffered a permanent deformation owing to the continued application of a large strain, the ends of the cylinders are prevented from abutting. This explains why under these conditions the cylinders cannot re-form, and no recovery of mechanical properties is observed.

7.4. The Microstructural Interpretation of Yielding
We have seen that at high strains the polystyrene cylinders are broken into shorter lengths and qualitatively this clearly reduces the modulus; this strain softening effect facilitates the formation of a neck. Beyond a

certain point no further strain softening can occur and the load again rises causing the neck to propagate along the sample.

We may now apply theoretical arguments to the high strain region, and attempt to explain quantitatively the interrelation between mechanical properties and microstructure.

7.4.1. The Shear-lag Theory for Short Rods

As we have seen already, and as will be presented in greater detail in Chapter 5, conventional composite theory is very successful at interpreting the mechanical properties in terms of microstructure for elastic deformations. We may attempt to extend this approach to the inelastic region by applying the shear-lag treatment for the calculation of mechanical properties of a short fibre reinforced composite.[25]

Consider a single cylinder of length l and radius r_0 in an elastic matrix. Close to a rod the matrix strain is non-uniform owing to the progressive transfer of stress to the rod. At a radial distance R from the centre of the rod, equal to the inter-rod spacing, the strain is considered to have a uniform value e. The system is shown in Fig. 21.

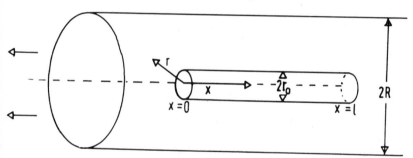

FIG. 21. The model used for the shear-lag treatment.[24]

The theory predicts that

$$\delta(x) = E_f e \left[1 - \frac{\cosh \beta \, (l/2 - x)}{\cosh \beta \, l/2} \right] \tag{7}$$

where x is the distance along the fibre from one end, $\delta(x)$ is the tensile stress in the fibre at x, E_f is the Young's modulus of the fibre, and

$$\beta = \frac{2G_m}{E_f \, r_0^2 \ln (R/r_0)}$$

where G_m is the shear modulus of the matrix material.

This analysis is not exact; in particular the matrix is assumed to carry no tensile load. A more exact analysis by Rosen[26] results in a different formulation for β:

$$\beta = \sqrt{\left(\frac{2G_m}{r_0 E_f(R-r_0)}\right)}$$

An exact treatment for the strain analysis was completed by Smith and Spencer[27] and their results are similar to those of Rosen.

Qualitatively, this shows that the tensile stress in the fibre increases from the ends to a maximum in the middle. However, only infinitely long cylinders can have the same strain at their centres as the matrix material.

Based on this model the Young's modulus of a simple composite would be

$$E_c = E_f V_f \left[1 - \frac{\tanh{(\beta l/2)}}{\beta l/2} \right] + E_m(1 - V_f) \tag{8}$$

where V_f is the volume fraction of the fibre and E_m is the Young's modulus of the matrix.

Let us consider eqn (7). The largest value of stress clearly occurs in the centre of the rod at $x = l/2$. Thus,

$$\sigma_{max} = E_f e \left[1 - \frac{1}{\cosh{(\beta l/2)}} \right] \tag{9}$$

The breaking stress of the rod σ_b is known from the stress–strain curve. When $\sigma_{max} = \sigma_b$ the rod will break. This situation will prevail for all rods beyond a certain length l_c depending on the value of the strain. This critical length, which is just long enough to be broken by the applied strain e, can be calculated from eqn (7) and is shown in Fig. 22. The value taken for σ_b is 2×10^7 Pa.

Since we know that rods longer than l_c will be broken by the applied strain, and rods of l_c, will just be broken, the lengths of rods remaining after an applied strain e will range from $l_c/2$ to l_c ($l_c/2$ is also plotted in Fig. 22).

Fitting dimensional values from the known microstructure and moduli from Ref. 20, Fig. 23, shows the predicted modulus as a function of remaining cylinder lengths for the S–B–S single-crystal material. The predicted modulus varies from that of the matrix at zero rod length to that of the perfect composite at infinite rod length.

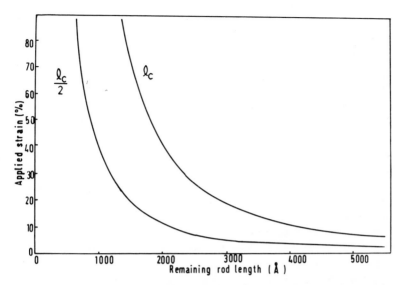

FIG. 22. Length of fractured rods remaining after a strain applied parallel to the rod direction, as predicted by shear-lag theory. The two curves show l_c and $l_c/2$, where l_c is the upper limit. The length of the remaining rods should lie between l_c and $l_c/2$.[24]

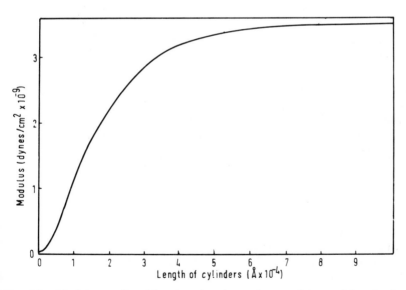

FIG. 23. Modulus predicted by shear-lag theory for remaining rod lengths 0–10^4 nm.

The shear-lag theory of composites requires a number of assumptions which are not entirely justified when applied to the present system.

The concept of shear as the mechanism for transferring stress to the cylinders may be inadequate. The shear modulus is a macroscopic parameter. It is known that in phase segregated block copolymers the rubbery phase is crosslinked effectively only at the phase boundary; there may be chain entanglements but little is known about them. An un-crosslinked material cannot be considered to transmit shear on a micro-scopic scale, although phenomenologically the analysis may give the right answer.

The strain is considered to be uniform and equal to the macroscopic sample strain at the surrounding set of cylinders; in the hexagonal lattice ($r = R$) this is manifestly not the case, and the justification for the approximation is not clear.

Another theoretical treatment has, however, addressed itself to these problems and is described in Ref. 24. Here a microscopic model of the deformation mechanism is used, taking into account the presence of the lattice of polystyrene cylinders. The treatment enables the calculation of the distribution of rod lengths as a function of applied macroscopic strain. The results predicted by this theory are broadly similar to those predicted by shear-lag theory but without unjustifiable assumptions.

7.4.2. Comparison with Electron Microscopy

It will be recalled that for material strained 80% the majority of the rods were found by electron microscopy to have broken to 70–110 nm, with a few very much shorter lengths down to 20 nm.

The theories above predict remaining rod lengths of 80–160 nm for this strain in excellent agreement with the observed rod lengths.

The observation of shorter rod lengths than predicted (down to 20 nm) may be caused by the presence of irregularities along the length of the rods serving to locally concentrate the stress, producing premature breakage. Indeed electron microscopy of ultra thin sections does seem to support this contention.[14]

7.5. The Mechanism of Yield and Necking

The models developed enable the quantitative mechanism of yield and necking to be predicted. Equations (7) and (9) give the remaining rod lengths as a function of strain and eqn (8) predicts modulus as a function of remaining rod length. Combining these gives the modulus as a function of strain, thereby enabling the calculation of the stress as a function of strain in the material during yield.

The stress initially decreases as the material starts to yield, then remains roughly constant as the strain in the yielding material increases to about 80%, as shown in Fig. 24. Beyond this strain, the stress starts to increase, exceeding that required to yield new material, which causes the neck to propagate.

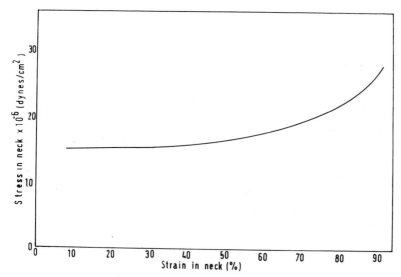

FIG. 24.   The stress in a sample as a function of strain in the yielding material.[24]

7.6. The Modulus of Yielded Material

A modulus of 3×10^6 Pa is predicted for the fully necked material, little more than the modulus of the matrix alone. Experimentally, however, the modulus is found to be 6×10^6 Pa; hence the polystyrene still has some reinforcing effect, which deserves a comment.

The fibre composite theories only consider the tensile load transferred to the cylinders by the matrix. Even when the reinforcement effect of this is very small, as for the necked material, the dispersed polystyrene phase still acts as a rigid filler in the rubbery polybutadiene. It is well known in rubber compounding technology that the inclusion of fillers can increase the modulus of a rubber considerably. Kerner[28] has predicted the effect of such inclusions:

$$E = E_{m}\left[1 + \frac{Q_2}{Q_1}\left(\frac{15(1-v)}{8-10v}\right)\right]$$

where E is the Young's modulus of the filled material, E_m the Young's modulus of the matrix phase, Q_1 and Q_2 the volume fractions of the matrix and dispersed phases, respectively, and v the Poisson's ratio of the matrix.

Applying this to the present case we predict $E = 5 \times 10^6$ Pa, now in reasonable agreement with the value of 6×10^6 Pa observed experimentally.

7.7. The Poisson's Ratio of the S–B–S Material

A rubber is generally considered incompressible and to have a Poisson's ratio (v) of 0·5 for unidirectional strain. However, with the present S–B–S material the styrene rods virtually constrain the butadiene matrix completely in one direction. As a consequence, from the expectation of the incompressibility of rubber, we would now anticipate an anisotropic Poisson's ratio. For strains applied perpendicular to the cylinders we expect v to be 1, whilst for strains applied along the direction of constraint (the cylinder direction) we expect v to be 0·5. Ascertaining this value, or any departure from it, in a rubbery material is of consequence for understanding the mechanical behaviour, deformation and underlying structure of the system.

There are three sources of information concerning the Poisson's ratio of single-crystal S—B—S material arising from our works.

1. The measurement of LAXD for strains applied perpendicular to the cylinders (Section 7.1) enables a precise determination of v from the variation of lattice parameters in the principal stress directions, a value of 0·74 being obtained, compared with the expectation of 1 for incompressibility.

2. The low strain composite analysis of Folkes and Arridge[20] (which will be developed further in Chapter 4) required a value of v of 0·37 for the butadiene for the best fit between observed and theoretical behaviour, compared with the expectation of 0·5.

3. The photographic method of Ref. 24 for the measurement of strain enables continuous assessment of v during a tensile test. With the applied stress along the rod direction the Poisson's ratio increases from 0·37 for elastic deformation to 0·48 for fully necked material, the latter being within experimental error of the value for an unconstrained incompressible rubber.

Thus we see that all three methods agree closely and give a value for the Poisson's ratio for elastic deformation which is smaller than the expected value, signifying that the material is compressible. We can safely assume that in our case this applies to the butadiene matrix. The compressibility may be caused by the constraint of the glassy polystyrene phase not permitting movement of the ends of polybutadiene molecules; any such constraint from Gaussian statistics is likely to increase the chain volume. In the necked material this constraint is removed by the disruption of the polystyrene phase, enabling the material to behave more like a rubber network.

Again all the above information and deductions therefrom, with all their further implications, could only become available through having the S–B–S material in the single-crystal form.

7.8. Conclusion

The elastic and inelastic deformation behaviour of single-crystal samples can be well accounted for by models based upon a highly anisotropic composite of regularly packed glassy cylinders in a rubbery matrix. This should thus serve as a basis for the interpretation of mechanical behaviour, up to and beyond yielding, of phase segregated block co-polymers under the more general circumstances of complex textures and stresses. Conversely, the present special samples also prove to be excellent test materials for displaying ideal composite behaviour in general. Some qualitatively new features of the block copolymer system (rod healing, 'crystallographic' cracking) are also directly revealed by virtue of the special nature of this material.

8. SWELLING BEHAVIOUR

Swelling is another property for which the single-crystal samples of S–B–S proved to be excellent models. When an agent is imbibed which swells the matrix but not the cylinders, clearly some predictable anisotropies are expected in the range of small swelling strains by analogy with the low strain mechanical behaviour, in both cases the resulting deformation of the components remaining within the elastic limits. At larger levels of swelling agent uptake, this behaviour is expected to become modified as the differential dimensional changes in the two components cannot be mutually accommodated without localized yield or fracture, again in broad analogy with mechanically induced deformation. However, at sufficiently large differential volume changes a further effect is expected to add to the above. Namely, it is known that the type of microphase morphology, i.e. whether we have spheres, cylinders or lamellae, is determined by the relative volume ratios as set out in Section 1. To this we now add that the influence of the volume ratios does not depend on whether the block ratio (in our case that of S and B) is changed, or whether the volume is changed, say increased, by the addition of a foreign agent such as a swelling medium. Accordingly, for an appropriate amount of matrix swelling a microphase transition, in the present case a cylinder to sphere transition, may be expected. It is with these anticipations that the swelling experiments on S–B–S single-crystals were undertaken. The effect of swelling was monitored by four types of measurement, not all equally exhaustively: dimensional changes, changes in the X-ray (low angle) patterns, changes in the birefringence and changes in the morphology.[29,30] The swelling itself was carried out in the vapour phase of a liquid which normally would have acted as a solvent for the pure homopolymer forming the matrix.

An example of the dimensional changes, as measured along both the perpendicular and parallel directions of the plug (or portions cut therefrom), is shown by Fig. 25 as a function of time. As shown by the concomitant measurement of weight, there occurs a gradually increasing amount of imbibition of the swelling agent, the additivity of the volumes being largely preserved (close correspondence between the curves defined by the dashed line and the circles). Confining our attention just to short times, corresponding to within about 20% total volume increase, we see that all the volume change corresponds to expansion in the perpendicular dimension and there is virtually no change in the parallel direction, this effect being reversible on removal and re-introduction of the swelling agent. Simultaneously, the low angle X-ray diffraction pattern, as recorded with the beam parallel to the plug (i.e. corresponding to the hexagonal pattern in Fig. 1(a)), revealed a closely 1:1 relation between changes in lattice spacings and changes in external dimension (Fig. 26). The picture which emerges is clearly that the matrix expands laterally with corresponding changes in the cylinder separation.

Beyond about 18% total volume change lateral expansion ceases and longitudinal expansion sets in at an accelerating rate, eventually exceeding the former. At the onset of the longitudinal change the expansion becomes irreversible; on repeated drying and swelling an excess longitudinal dimensional increase remains. The explanation is again straightforward: the matrix attempts to expand isotropically but is constrained by the unswollen cylinders. However, under the increasing forces acting along the longitudinal direction the glassy cylinders should eventually break up causing the effect observed.

In the irreversible range the X-ray patterns become more difficult to

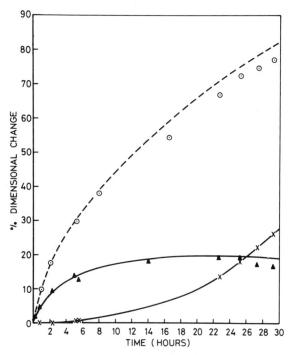

FIG. 25. Dimensional changes of single-crystal samples of Kraton 102 during swelling in decane: ×, parallel to extrusion direction; ▲, perpendicular to extrusion direction. Percentage volume change: ⊙, from dimensions; –––, from weight uptake of solvent.[29]

interpret. A hexagonal pattern of reflections was dominant along *all* directions, i.e. not only with the beam parallel but also perpendicular to the plug axis (where the unswollen plugs give rise to patterns as in Fig. 1(b)). This on its own indicates that a periodicity *along* the original cylinders has arisen, consistent with the postulated breaking up of the cylinders. However, the various essentially hexagonal patterns could not be made mutually consistent by any simple model. Hexagonal patterns obtained with the beam perpendicular to the plug axis clearly suggest a spherical microphase structure into which the cylindrical structure has transformed. Nevertheless all attempts to identify a single-crystal of spherical microphase by any of the simple sphere packings were unsuccessful. Electron microscopy on sections of swollen and subsequently dried samples (drying did not significantly change the X-ray pattern in

A. KELLER AND J. A. ODELL

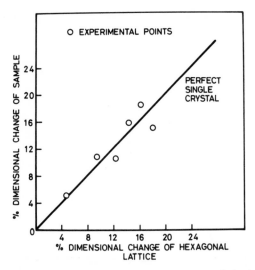

FIG. 26. Change of sample dimensions versus change of single-crystal lattice parameter during swelling of Kraton 102 as determined by low angle X-ray diffraction.[29]

this irreversible range) revealed a broadly hexagonal pattern of circles (somewhat as in Fig. 2) even in longitudinal sections (which gave images as in Fig. 3 in the unswollen case). This again reveals that the cylinders have broken up, and at first sight suggests that they have become spheres. Again, however, full consistency with a single-crystal constituted by spheres could not be achieved. On the other hand complex features could be located suggesting a complex highly symmetrical network structure of short cylinders displaying in most views a dominance of circular cross-sections (Fig. 27). At this stage the work, for the first time in this whole group of studies on single-crystal samples, has become inconclusive. Accordingly, this swelling work indicates more complex microphase structures than that corresponding to spheres for high matrix–particle ratios, contrary to basic expectations in this field. Nevertheless, the observations are at least consistent with complex networks of microphases identified by Luzzati et al. in analogous lipid systems by X-ray diffraction alone.[8] The following up of this parallel is clearly a challenge (a note on which we closed our last publication on this subject[30]) and to our knowledge has not been taken up anywhere since. These findings are challenging the generality, if not the existence, of

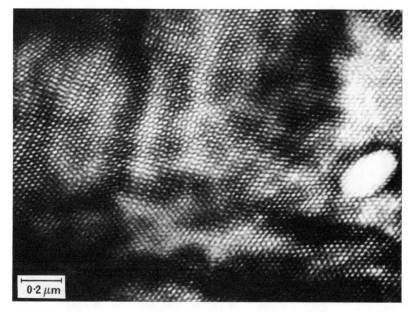

0·2 μm

FIG. 27. Electron micrograph of Kraton 102 swollen in hexane vapour, sectioned parallel to the cylinder direction. The continuous cylinders of Fig. 3 are disrupted, resulting in a complex microstructure (see text).[30]

the picture of spherical arrays of microphases at highly disparate component ratios, the picture taken for granted in the field of block copolymers, notwithstanding the different microphase structures confirmed under analogous circumstances in the lipid systems by much more powerful evidence. The recognition, if not the solution, of this problem was made possible on our part by using single-crystal samples of the cylindrical microphase type for the swelling experiments, with all the additional information this is capable of conveying.

Changes in the birefringence on swelling deserve a mention. As we have shown the birefringence of the unswollen sample is virtually entirely due to form birefringence. By eqn (1) this is influenced by the refractive indices of the two phases. It follows that an imbibition of the swelling agent should cause predictable changes in the overall birefringence by its effect on the form birefringence alone. Indeed, reasonable correspondence between measurement and theory has been found (Fig. 28). Complete agreement is not expected in view of the stresses which are anticipated in

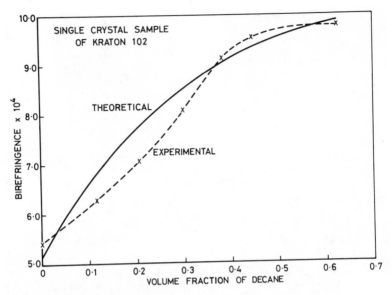

FIG. 28. Comparison of the experimental and theoretical dependence of the form birefringence of single-crystal samples of Kraton 102 on volume fraction of swelling agent (decane).[30]

the system as a result of swelling. In view of this it is remarkable that the results conform to expectations from form birefringence alone as well as they do.

9. OTHER DOMAIN PACKING SYSTEMS

Spherical domains of the dispersed phase have been previously mentiond and visibly demonstrated by ourselves and Pedemonte and co-workers,[21] in systems where the volume fraction of the dispersed phase is lower than that giving rise to cylindrical structures. Cubic packing has been observed with electron microscopy.[21] However, it has not proved possible to obtain macroscopic single-crystal samples, and further discussion is not helpful for the present purpose.

More significantly, with a larger volume fraction of the dispersed phase, a system of alternating lamellae can be achieved with appropriate preparation methods. This can be considered as a one-dimensional, single-crystal, structure. Macroscopic samples displayed the required mechanical anisotropy, even if quantitative values did not compare as closely with anticipations from properties of the components as in the case of cylindrical morphology,[31] because the perfection of the structure was not as good. Nevertheless, the anisotropy was very marked in the swelling behaviour, conforming well with expectations.[32]

In contrast to the cylindrical morphology, birefringence measurements as a function of solvent uptake revealed some *intrinsic* chain orientation in the microphases perpendicular to the lamellae. More perfect lamellar structures have been obtained by Terrisse:[33] X-ray patterns show up to 10 orders of reflection, this proving particularly helpful to the understanding of the interface problem (see below).

By further increasing the styrene content the butadiene becomes the dispersed phase, resulting in structures of spheres and cylinders in a styrene matrix. A macroscopic single-crystal texture of butadiene cylinders has in fact been obtained from a material of 70% styrene content. This was verified by low angle X-ray diffraction and transmission electron microscopy as previously applied to systems where the styrene forms the dispersed phase, Fig. 29.[34] This inverse system was glassy in

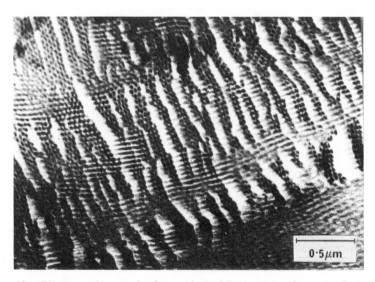

FIG. 29. Electron micrograph of a cracked oblique section from a polystyrene rich S–B plug enabling a perspective view of the polybutadiene cylinders which appear dark.[34]

character and thus could serve as a model for rubber-reinforced polystyrene; this line, however, has not been pursued further.

10. THE MICROPHASE INTERFACE

In the case of cylindrical morphology the optical and mechanical evidence is interpreted in terms of a well-defined two-component structure. There was no need to invoke (or consider) the existence of an interfacial zone of partial mixing between the two phases. Nevertheless, the existence of such a layer is a central issue in the study of phase segregation in block copolymers, with thermodynamic theories of phase separation predicting an interfacial layer which can consist of a sizeable fraction of the total composition, increasing with higher molecular wieght.[9,35,36]

The single-crystal nature of the present materials results in many orders of low angle X-ray reflections. In the case of lamellar systems the diffraction has been Fourier analysed and information on the interface obtained. Terrisse concludes that the interfacial thickness is slight,[33] but the work is restricted to comparatively low molecular weights, which would not be expected to yield thick interfacial regions.[35] Other diffraction work by Kawai and co-workers[37] on less perfectly oriented material, but using a range of molecular weights, does support the contention that the size of an imperfectly phase-segregated layer increases with increasing molecular weight. Mechanical measurements of loss moduli on randomly oriented samples[38] reach broadly similar conclusions.

11. SOME FORWARD-LOOKING ASPECTS

The present single-crystal materials give an opportunity for testing composite theories in general, with a structure close to the ideal composite, in terms of perfect orientation, near-infinite cylinders of the dispersed phase, molecular connectedness and very large differences in mechanical properties between the components.

As will be detailed in subsequent chapters this has enabled a number of further advances.

First, existing composite theories can be tested and the theoretical predictions discriminated between. Secondly, the results have served to highlight certain general effects in the measurement of mechanical properties in highly anisotropic systems (the importance of end-effects and the application of St. Venant's principle) with wider implications for mechanical testing (see Chapter 4).

Finally, current work in this field is proving to be of practical consequence: the idealized single-crystal results can be applied to aid the understanding of 'real-life' fabricated articles, the properties of which can now be interpreted in terms of microstructure in an explicit manner (see Chapter 4).

REFERENCES

1. Keller, A., Dlugosz, J., Folkes, M. J., Pedemonte, E., Scalisi, F. P. and Willmouth, F. M. (1971). *J. Physique Colloque C5a*, Suppl. 10, **32**, 295.
2. Folkes, M. J. and Keller, A. (1973). In: *Physics of Glassy Polymers*, (Ed. R. N. Haward), Applied Science Publishers Ltd, London.
3. Folkes, M. J. and Keller, A. (1973). In: *Block and Graft Copolymers*, (Ed. J. Burke and V. Weiss) Proceedings 19th Sagamore Army Materials Research Conf., Syracuse University Press.
4. Folkes, M. J. and Nazockdast, H. (1985). To be published.
5. Arridge, R. G. C. and Folkes, M. J. Chapter 4 in this book.
6. Allport, D. G. and Janes, W. H. (Eds), (1973). *Block Copolymers*, Wiley, New York.
7. Aggarwal S. L. (Ed.), (1970). *Block Copolymers*, Plenum Press, New York.
8. Luzzati, V., Tardieu, A., and Gulik-Krzywicki, T. (1968). *Nature*, **218**, 1031.
9. Helfand, E. and Wassermann, Z. R. (1978). *Macromolecules*, **11**, 960.
10. Keller, A., Pedemonte, E. and Willmouth, F. M. (1970). *Nature*, **255**, 538.
11. Keller, A., Pedemonte, E. and Willmouth, F. M. (1970). *Kolloid Z.u.Z.*, *Polymere*, **238**, 385.
12. Folkes, M. J., Keller, A. and Scalisi, F. P. (1973). *Kolloid Z.u.Z. Polymere*, **251**, 1.
13. Dlugosz, J., Keller, A. and Pedemonte, E. (1970). *Kolloid Z.u.Z. Polymere*, **242**, 1125.
14. Odell, J. A., Dlugosz, J. and Keller, A. (1976). *J. Polym. Sci., Polym. Phys. Ed.*, **14**, 847.
15. Kato, K. (1966). *J. Polym. Sci.*, **B4**, 35.
16. Folkes, M. J. and Keller, A. (1971). *Polymer*, **12**, 222.
17. Pechhold, W. R. and Grossmann, H. P. (1979). *Faraday Disc. Chem. Soc.*, **68**, 58.
18. Folkes, M. J., Keller, A. and Scalisi, F. P. (1971). *Polymer*, **12**, 793.
19. Nye, J. F. (1957). *Physical Properties of Crystals*, Oxford Clarendon Press, Oxford.
20. Folkes, M. J. and Arridge, R. G. C. (1972). *J. Phys., D*, **5**, 344; or (1975). *J. Phys., D*, **8**, 1053.
21. Turturro, A., Bianchi, U., Pedemonte, E. and Ravetta, P. (1972). *Chimica Ind. Milano*, **54**, 782.

22. Pedemonte, E., Cartasegra, S. and Turturro, A. (1974). *Chimica Ind. Milano*, **56**, 3.
23. Pedemonte, E., Turturro, A. and Dondero, G. (1974). *Brit. Polym. J.*, **6**, 277.
24. Odell, J. A. and Keller, A. (1977). *Polym. Eng. Sci.*, **17**, 544.
25. Cox, H. L. (1952). *Brit. J. Appl. Phys.*, **3**, 72.
26. Rosen, B. W. (1965). 'Fibre composite materials', Am. Soc. Metals, p. 37.
27. Smith, G. E. and Spencer, A. J. M. (1970). *J. Mech. Phys. Solids*, **18**, 81.
28. Kerner, E. H. (1956). *Proc. Phys. Soc.*, **B69**, 808.
29. Folkes, M. J. and Keller, A. (1976). *J. Polymer Sci., Polym. Phys. Ed.*, **14**, 833.
30. Folkes, M. J., Keller, A. and Odell, J. A. (1976). *J. Polym. Sci., Polym. Phys. Ed.*, **14**, 847.
31. Folkes, M. J. and Keller, A., unpublished work.
32. Folkes, M. J. and Keller, A. (1976). *J. Polym. Sci., Polym. Phys. Ed.*, **14**, 833.
33. Terrisse, J. (1974). Thesis, University of Strasbourg.
34. Folkes, M. J., Elloway, H. F., O'Driscoll, G. M. M. and Keller, A. (1977). *Polymer*, **18**, 960.
35. Meier, D. J. (1974). *Polym. Preprints*, **5**, 171.
36. Krömer, H., Hoffmann, M. and Kämpf, G. (1970). *Ber. Bunsenges Physik, Chem.*, **74**, 859.
37. Hashimoto, T., Nagatoshi, K., Todo, A., Hasegawa, H. and Kawai, H. (1974). *Macromolecules*, **7**, 364.
38. Kraus, G. and Rollmann, K. W. (1976). *J. Polym. Sci., Polym. Phys. Ed.*, **14**, 1133.

CHAPTER 3

Melt Flow Properties of Block Copolymers

J. LYNGAAE-JØRGENSEN

*Instituttet for Kemiindustri, The Technical University of Denmark,
Lyngby, Denmark*

1. INTRODUCTION

The appearance of commercial large scale production of a number of
different block copolymer systems has stimulated interest in new applications and new products; this necessitates a better understanding and
knowledge of the processing behaviour and, therefore, also a better
understanding of the rheological properties of block copolymers in the
melt state.

Much emphasis has recently been placed on the possibility of monitor-
ing material properties by 'structuring' non-compatible blends of different polymer systems, especially polymer blends. Structuring is used in
the sense of governing the 'morphology' of two-phase systems.

A number of monographs on block copolymers have been published
(see Chapter 1) and most of these refer to flow behaviour. The flow
behaviour of two-phase blends has been reviewed extensively in the
literature, e.g. by van Oene[1] and quite recently in the monograph by
Han.[2]

To differentiate the treatment discussed in this chapter from others,
emphasis will be placed on the description/evaluation of flow units, flow
mechanisms and rheological properties during flow of block copolymers.
Here stress is placed on flow behaviour in simple flows and the treatment
is furthermore confined to studies in the melt flow–fluid region of block
copolymers with non-compatible blocks (e.g. temperatures above the
glass transition temperature of PS). Solid-state behaviour is not considered in this chapter. Finally, the behaviour of undiluted block copolymers will be given priority.

The block copolymers discussed in this chapter are thermoplastic and the molecular structure is considered to be a given basic property of a material, i.e. unless otherwise specified, the single polymer molecule is assumed not to undergo structural change during rheological measurements.

2. THE STRUCTURE OF BLOCK COPOLYMER, FLOW MECHANISM AND FLOW UNIT

The properties of a block copolymer are determined by a number of basic structural factors. A non-exhaustive survey of these factors is schematically presented for block copolymers built of two comonomers as follows:

1. Molecular structure — repetition units
                       — distribution of sequence length in blocks
                       — number of blocks per polymer molecule; distribution of number of blocks per polymer molecule
                       — molecular weight distribution, branch structures, etc.
2. Microstructure   — multiphase morphology
                       — crystallinity
                       — orientation effects
                       — compounding ingredients, etc.
3. Macrostructure   — geometrical form of the material

The microstructure of non-compatible block copolymers depends very much on the molecular structure. The two-phase microstructure can be considered as comprising spherical or cylindrical domains in a continuous matrix or as lamellar domains (see Chapter 1). The domain form depends primarily on the volume ratio of the comonomers.

Experimental evidence (Section 4) shows that if a block copolymer with a two-phase structure is heated and deformed it will experience changes in the microstructure. Such structural changes depend on time and deformation. This means that two-phase (multi-phase) block copolymers behave as time-dependent viscoelastic fluids. The structure breakdown and the establishment of a steady-state flow structure presents both experimental and theoretical difficulties, and it is fair to say that the analysis of the flow of multiphase polymer systems has not advanced very far, leaving many unanswered questions.

Arnold and Meier[3] interpret the structure breakdown for an ABA lock copolymer in simple flow as illustrated roughly in Fig. 1. ccording to this interpretation the original structure (e.g. dispersed pheres of phase A in a continuous phase of B consituting a three-mensional network held together by the A domains) is, when experiencing increasing rates of deformation, gradually broken down to ructures consisting of flowing aggregates behaving as star-shaped mol-

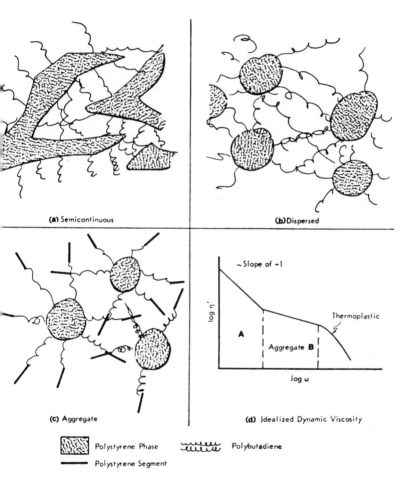

(a) Semicontinuous   (b) Dispersed

(c) Aggregate   (d) Idealized Dynamic Viscosity

Polystyrene Phase   Polybutadiene

Polystyrene Segment

G. 1.   Idealized structures. Reproduced from Ref. 3 by courtesy of the publisher, John Wiley and Sons, Inc.

ecules. Finally, at sufficiently high deformation rates, a monomolecular melt state will possibly be established.

In turning to the important question of how the microstructure is actually deformed during deformation and flow and which criteria govern the stability of the structure of block copolymers during flow, the limit of knowledge available at present has already been crossed. Very few systematic investigations in this sphere of problems have been published, and no definite conclusions may be reached.

Difficulties in analysing the rheological data for block copolymers reflect very often the fact that their rheological properties are influenced by complex structural factors, such as domain size and shape, interface size and structure, and volume fraction of the dispersed phase during flow. Under these circumstances it is recommended that studies of their flow behaviour are made in the simplest possible flow fields in order to interprete rheological data for multiphase systems. Thus, for example, cone and plate studies in shear flow[10] seem to offer distinct advantages because of the constant shear rate and shear stress in this geometry.

At present, process analyses may be based on a rheological equation for viscoelastic materials but are often based on either an assumption of power law melt behaviour or (semi)-empirical correlations between process conditions and rheological functions measured in simple flows. Particularly in the case of melts with complex flow units, a simplified approach is necessary.

For the copolymer melts containing complex melt structures, flow data are divided into steady-state equilibrium data and non-steady-state data. A short section defining the simple flows mostly used to study the flow behaviour of block copolymers is given in Section 3.1. This chapter is based on data measured in these simple flows.

2.1. Polymer Blends, Two-phase Flow

Some guidelines for the flow of block copolymers may eventually be drawn from studies on polymer blends. A very brief survey on structure changes associated with the flow of blends is given below.

The flow of Newtonian droplets in Newtonian media has been extensively covered in the literature. Almost complete solutions have been obtained by Taylor[104] and Cox.[105] These authors have solved the dynamic equations with Newtonian liquids under the following boundary conditions: the normal and tangential velocity components are continuous through the interfaces, the tangential components of the stresses are continuous, and the difference in the normal components is balanced by interfacial tension forces on the deformed surface. The droplet break-up criterion used by Taylor was that a drop will burst when the maximum difference in the components of stress tending to disrupt the drop exceeds the force, due to surface tension, which tends to hold it together. The subject was reviewed by Goldsmith and Mason.[106]

Relatively few theoretical and experimental studies dealing with the deformability and break-up of viscoelastic droplets suspended in another viscoelastic liquid have been published. Van Oene[4] has advanced a theory for the formation of droplets from a mixture of two viscoelastic fluids, based on thermodynamic arguments. The main idea is that the free energy of each phase changes with the deformation of the liquids, namely with the first normal stress difference. If the normal stress difference is different in the two phases, van Oene argues that droplet formation may be followed by a negative change in free energy. Han and Funatsu[5] have proposed the demarcation of stable from unstable regions, using the concepts from the analysis of Newtonian systems. Thus, a stable region is empirically located in a delineation of the Weber number

$$\frac{\eta_B \dot{\gamma}_c^{\,a}}{\sigma} \text{ against } \eta_A/\eta_B$$

in which η_A and η_B are viscosities of the droplet phase and suspending medium, respectively, $\dot{\gamma}_c$ is the critical value of apparent shear rate for droplet break-up, a is the droplet radius, and σ is interfacial tension.

Chin and Han[6,7] have recently analysed the flow behaviour of a two-phase viscoelastic system at the inlet region to a capillary and performed a stability analysis of an elongated droplet in the fully developed shear flow region in a capillary. A very brief review of some of the most important experimental results for the flow of dispersed two-phase polymer blends during capillary flow is given below:[1,2,4-9]

1. The discrete droplets are deformed during the inlet to the capillary into fibrils or elongated droplets. The amount of deformation depends on the relative viscosity, the droplet size and the volumetric flow rate.

2. The fibrils run through a recoil or stress relaxation zone where the droplets change form. This zone is well inside the capillary. If break-up takes place, it very often takes place in this zone. If the ratio between domain viscosity and the viscosity of the continuous phase is large, the domains are stable. Small domains are more stable than large ones and shear rate influences stability.

3. Quite recently, Tsebrenko, *et al.*[107] observed that very long fibrils could only be obtained at a viscosity ratio close to one.

4. Han[2] presents an extensive treatment of two-phase flow including the droplet break-up problem.

3. EXPERIMENTAL MEASURING DATA AND EXAMINATION OF THE FLOW BEHAVIOUR OF BLOCK COPOLYMERS

As mentioned above the evaluation of flow behaviour is confined to a few well-defined simple flow cases in order to make an attempt to interpret the anomalies of these thermorheologically complex systems[11] easier. The definitions of the applied rheological functions are outlined briefly in the following sections. Comprehensive treatments may be found in many textbooks on rheology.[1,11–13]

3.1. Rheological Functions Measured in Simple Flow Fields

3.1.1. The Steady Shearing Flow Field
Consider a flow geometry consisting of two infinitely long, parallel planes, forming a narrow gap filled with a fluid. The plates are separated by distance H, which is very small compared with the width, W, of the plane (i.e. $H \ll W$), as shown schematically in Fig. 2. The velocity field of the laminar flow at steady-state between the two planes is given by

$$v_x = f(y), \qquad v_z = v_y = 0$$

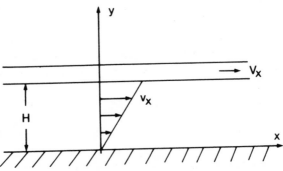

FIG. 2.   Schematic diagram of simple shearing flow.

Referring to Fig. 2 the lower plane is stationary (i.e. $v_x(0)=0$) and the upper plane moves in the x direction at a constant speed, v_x, i.e. $v_x(H)=V_x$. We then have a velocity gradient, dv_x/dy which is constant, i.e.

$$\dot{y}=dv_x/dy=\text{const.} \tag{3.1}$$

where $\dot{y}$ is the shear rate.

In steady simple shearing flow the stress tensor may be written

$$\tau = \begin{Bmatrix} \tau_{xx} & \tau_{xy} & 0 \\ \tau_{yx} & \tau_{yy} & 0 \\ 0 & 0 & \tau_{zz} \end{Bmatrix} \tag{3.2}$$

where the extra stress tensor $\tau = S + P \cdot \delta$ (S is the total stress tensor, P is the hydrostatic pressure and δ is the unit tensor. The subscript x denotes the direction of flow in Fig. 2, y denotes the direction perpendicular to the flow and z denotes the neutral direction. $\tau_{xy}=\tau_{yx}$ is the shear stress and τ_{xx}, τ_{yy}, τ_{zz} are normal stresses.

Three material functions are sufficient to describe the stress–shear rate dependence in simple shear flow. These are normally defined by

$$\tau_{yx}=\eta(\dot{y})\dot{y}, \quad \tau_{xx}-\tau_{yy}=\psi_1(\dot{y})\dot{y}^2$$

and

$$\tau_{yy}-\tau_{zz}=\psi_2(\dot{y})\dot{y}^2$$

where $\eta(\dot{y})$ is the viscosity function and $\psi_1(\dot{y})$ and $\psi_2(\dot{y})$ are the first and second normal stress coefficients, respectively. The experimental techniques most used for the determination of the rheological functions in simple shear involve the use of capillary viscometers (Poiseuille flow), coaxial cylinder viscometers (Couette flow) and cone and plate viscometers.[10,94,95]

3.1.2. Dynamic Data (Oscillatory Shear Flow)

The displacement gradient dx/dy (Fig. 2) is known as the shear strain and given the symbol γ

$$\gamma = \frac{dx}{dy} = \text{shear strain (dimensionless)}$$

The time rate of change of shear strain $\dot{\gamma}$ (the dot is Newton's notation for the time derivative) is the shear rate given in eqn (3.1).

If a linear viscoelastic fluid is subjected to a sinusoidally varying shear strain, γ in Fig. 2, the measured stress, τ will also vary sinusoidally but normally out of phase with the strain or

$$\gamma = \gamma_0 \cos \omega t \qquad (3.3)$$

$$\tau = \tau_0 \cos(\omega t + \delta) \qquad (3.4)$$

where γ_0 and τ_0 are the maximum values of strain and stress, respectively, ω is the circular frequency, t is time and δ is the phase angle. The analysis of dynamic data is most easily performed by introducing complex quantities. A complex shear strain and shear stress may be defined as follows:

$$\gamma^* = \gamma_0(\cos \omega t + i \sin \omega t) = \gamma_0 \exp(i\omega t) = \gamma' + i\gamma''$$

where

$$|\gamma^*| = \gamma_0 \text{ and } \gamma = Re\,\gamma^* \qquad (3.5)$$

and

$$\tau^* = \tau_0(\cos(\omega t + \delta) + i(\sin \omega t + \delta)) = \tau_0 \exp\,(i(\omega t + \delta))$$

where

$$|\tau^*| = \tau_0 \text{ and } \tau = Re\,\tau^* \qquad (3.6)$$

Using the definitions given in eqns (3.5) and (3.6) a complex modulus,

$$G^* \equiv \frac{\tau^*}{\gamma^*} = G' + iG'' \qquad (3.7)$$

and a complex viscosity

$$\eta^* = \frac{\tau^*}{\gamma^*} = \frac{\tau^*}{i\omega\gamma^*} = \eta' - i\eta'' \qquad (3.8)$$

both of rheological significance, may be defined.
Thus

$$G^* = \frac{\tau_0 \exp(i\omega t + \delta)}{\gamma_0 \exp(i\omega t)} = \frac{\tau_0}{\gamma_0} \exp(i\delta) = \frac{\tau_0}{\gamma_0}(\cos \delta + i \sin \delta) \text{ or } G' = \frac{\tau_0}{\gamma_0} \cos \delta$$

and

$$G'' = \frac{\tau_0}{\gamma_0} \sin \delta$$

ıd

$$\eta^* = \frac{\tau^*}{i\omega\gamma^*} = \frac{i}{i}\left(\frac{G'}{i\omega} + \frac{iG''}{i\omega}\right) = \frac{G''}{\omega} - \frac{iG'}{\omega}$$

$$\eta' = \frac{G''}{\omega} \text{ and } \eta'' = \frac{G'}{\omega}$$

(ω) is the dynamic viscosity and $G'(\omega)$ is the storage modulus. G' an η'' n be shown to be associated with energy storage and release. $G''(\omega)$ is lled the loss modulus. G'' and η' are associated with the dissipation of ergy as heat. A large number of instruments can be used to measure these rheologi-l functions.[10,94,95]

1.3. Uniaxial Elongational Flow

flow field of great practical importance is uniaxially extensional flow at may be found in such polymer fabrication processes as fibre inning and processing, involving converging entry flows.[93] For un-xial stretching (e.g. fibre spinning) the velocity field is given as

$$v_x = f(x), \quad \frac{\partial v_z}{\partial x} + \frac{\partial v_y}{\partial y} + \frac{\partial v_z}{\partial z} = 0 \tag{3.9}$$

quation (3.9) assumes a flat velocity profile in the directions per-ndicular to the flow direction. For such a flow field, the rate-of-strain isor $\dot{\gamma}$ is

$$\dot{\gamma} = \left\{ \begin{array}{ccc} \dot{\gamma}_E & 0 & 0 \\ 0 & -\dot{\gamma}_E/2 & 0 \\ 0 & 0 & -\dot{\gamma}_E/2 \end{array} \right\}$$

which $\dot{\gamma}_E$ is called the elongation rate, defined as

$$\dot{\gamma}_E = dv_x/dx \tag{3.10}$$

For uniaxial elongation flow, the elongational viscosity η_E may be fined by the ratio of tensile stress S_{xx} and rate of elongation (or ongation rate) $\dot{\gamma}_E$,

$$\eta_E = S_{xx}/\dot{\gamma}_E \tag{3.11}$$

If the surfaces transverse to the direction of principal elongation (i.e. the direction of stretching) are unconstrained, then

$$S_{yy} = S_{zz} = 0 \qquad (3.12)$$

and eqn (3.11) may be rewritten

$$\eta_E = (\tau_{xx} - \tau_{yy})/\dot{\gamma}_E$$

Over the last few years a large number of measuring instruments have been developed which measure η_E.[94,95]

3.2. Analysis of Flow Data — Structure Evaluation

Apparently, the most frequently applied 'method' used to assess the influence of 'structure' on properties is to study the rheological properties of the block copolymers directly and compare these properties with the properties 'expected' for a homogeneous melt where the single polymer molecules constitute the flow units (called a monomolecular melt). The material properties measured by application of large deformations are generally strong functions of the molecular and micro structure of the materials. In this treatment, this technique will be tried in connection with a rough testing of the hypothesis presented in Section 5. For monomolecular melts of linear polymer molecules in simple flows the following master curve relations were found to be in reasonable accord with experimental evidence.[14]

Simple shearing flow

$$\eta/\eta_0 = F_1(\lambda\dot{\gamma}) \text{ or } \eta/\eta_0 = F_1'(\tau/K_1) \qquad (3.13)$$

where

$$\lambda = \frac{\eta_0 M_c H\rho}{c^2 RT} \cdot a \text{ and } K_1 = \frac{c^2 RT}{M_c H\rho}$$

and

$$\frac{\tau_{xx} - \tau_{yy}}{K_1} = F_2(\lambda\dot{\gamma}) \text{ or } \frac{\tau_{xx} - \tau_{yy}}{K_1} = F_2'\left(\frac{\tau}{K_1}\right) \qquad (3.14)$$

$$\frac{\tau_{yy} - \tau_{zz}}{K_1} = F_3(\lambda\dot{\gamma}) \text{ or } \frac{\tau_{yy} - \tau_{zz}}{K_1} = F_3'\left(\frac{\tau}{K_1}\right) \qquad (3.15)$$

where η is the viscosity at temperature T, η_0 is the zero shear viscosity, M_c is twice the molecular weight between entanglements, H is the heterogeneity index: $\bar{M}_w/\bar{M}_n$, ρ is the density of pure polymer, c is the polymer concentration and a is a constant, which has to be equal to 3 if maximum relaxation times from stress relaxation measurements are used for 'calibration'.

Dynamic measurements

$$\frac{\eta'}{\eta'_0} = F_4(2\lambda\dot{\omega}) \tag{3.16}$$

A necessary condition for application of this approach is that the zero shear viscosity of a monomolecular melt state can be measured or at least estimated. At present an equation correlating the zero shear viscosity of a monomolecular melt state with measurable quantities, e.g. molecular weight and composition of block copolymers, (a mixing rule) is not at hand. This considerably restricts the use of a master curve approach.

Nevertheless, whenever possible these 'master curves' will be used for comparison with experimental data in order to test hypotheses concerning transitions to monomolecular melt states in the sense that agreement between master curve and experimental data may be taken as a necessary condition for homogeneity.

Unfortunately, agreement with such a master curve is not a sufficient condition for homogeneity, since data for non-compatible blends may comply with such master curves.[15] The establishment of anomalous rheological behaviour of block copolymers compared with the expected behaviour of monomolecular melts has been used as a diagnostic tool and has been interpreted in terms of structural changes, as discussed in Section 4. Because of experimental and theoretical difficulties, most evaluations of structural changes during flow reported in the literature are performed indirectly by observing the structure immediately before and after a flow experiment is performed and use is often made of quenching in order to freeze the structure. In these cases, all the methods generally applicable to the analysis of blends are, in principle, available. Included are electron microscopy, optical methods (light diffraction, birefringence, light scattering), low angle X-ray and different measurements of the T_g of the material (see Chapters 1 and 2). Unfortunately, direct structure observations during flow are few in number. Many of the measuring techniques mentioned above could in principle be applied directly on a melt during flow. Most directly applicable are different types of rheo-optical methods[16] such as small angle light scattering, dichroism, X-ray diffraction, fluorescence and turbidity measurements.[17]

Flow birefringence measurements have been applied rather extensively in the study of monomolecular polymer melts (e.g. Wales[18]) but few investigations on block copolymers (with the exception of homogeneous solutions[19,20]) have been published.

Birefringence or perhaps turbidity measurements are, in this author's opinion, the most promising techniques. Only a brief treatment of birefringence measurements will be given in this chapter.

The optical properties of a material may be described by three refractive indexes n_i, n_j, n_k measured along three mutually perpendicular axes in the material. When the three refractive indexes are equal, the material is optically isotropic.

Birefringence, Δn, is defined as the difference between refractive indexes in two of the orthogonal directions, e.g. i, j, as measured with polarized light, i.e.

$$\Delta n = n_i - n_j$$

When an isotropic polymer is oriented, e.g. by drawing, the refractive index parallel to the direction of draw, $n_{\parallel}$, is no longer equal to the refractive index perpendicular to this direction, $n_{\perp}$, and consequently the polymer displays birefringence $\Delta n = n_{\parallel} - n_{\perp}$. The extent and sign of the birefringence depends on the perfection of orientation and the optical anisotropy of the units making up the polymer. Birefringence measurements give an average orientation index. In multi-component systems it may be difficult to interpret the result, since the birefringence is influenced by at least two contributions, namely, the orientation birefringence $(\Delta n)_k$ and the form birefringence $(\Delta n)_f$. One can write:

$$\Delta n = \sum_i \phi_i (\Delta n)_k + (\Delta n)_f$$

The technique presents some unsolved problems concerning the interpretation of experimental data. Specifically, an unambiguous evaluation of the contributions to the birefringence of the form birefringence and orientation birefringence in a flowing system needs to be demonstrated in analogy with the work of Folkes and Keller[21] on an extruded styrene/butadiene block copolymer sample.

4. EXPERIMENTAL INVESTIGATIONS ON BLOCK COPOLYMERS

The rheological behaviour of non-compatible AB, ABA, $A_n B_n$ block copolymers is considerably different from that of homopolymers as well as random copolymers. The block copolymers are reported to have higher melt viscosities (at low rates of deformation) than random copolymers having the same composition and weight average molecular weight. It seems to be generally observed that block copolymers and sometimes also random copolymers do not approach a constant viscosity at low shear rates in simple steady-state shear flow. Abrupt changes of slopes in the flow curves of block copolymers have been observed by many research groups.

Figure 3 depicts possible shear viscosity–shear rate curves for given three-block copolymer samples measured at steady state. The full drawn curve represent the normally observed curve type for a monomolecular melt. Deviations may be observed at high shear rates/high shear stresses, e.g. caused by melt fracture. The dashed lines represent possible curves

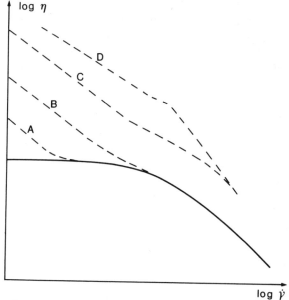

FIG. 3.   Possible steady-state shear viscosity–shear rate curves for three-block copolymers.

for block copolymers. Curves A and B show non-Newtonian behaviour (structure breakdown) at low shear rates and coincidence with the viscosity curve for monomolecular melts at high shear rates. Curves C and D deviate from the viscosity–shear rate curve for a monomolecular melt at all measurable shear rates.

There is no general consensus concerning the 'flow units' and flow mechanisms. However, it seems clear that the complex melt flow behaviour can be attributed to the existence of the two-phase domain structure and domain structure changes as a function of the variables which determines the flow.

By far the best investigated types of block copolymer are those of styrene and butadiene, and here again the three-block S–B–S-type has received most attention. This is an obvious consequence of the fact that these materials were the first 'large' scale block copolymers commercially available.

It is also for an S–B–S block copolymer that the most comprehensive investigation of different melt flow properties has been performed. In particular, the IUPAC investigation reported in Ref. 22 should be mentioned in this context. The data reported in the IUPAC Report[22] represent co-operative data obtained by 14 major research laboratories. These extensive studies have been performed on Cariflex TR-1102, an S–B–S-type manufactured by Shell. The block lengths in molecular weight units are 11 000–56 000–11 000 and the styrene content is about 28% by weight ($\bar{M}_w/\bar{M}_n = 1 \cdot 2$). Data for this sample are reported in Refs 22, 23 and 24.

An examination of the literature, which is the subject of the following sections, will of course reflect the fact that most accessible data are reported for S–B–S-types. The literature basis is biased in this respect and generalizations must be exercised with caution. An attempt has been made to treat the accessible literature data according to measuring method and within each method to treat separately the effect of rate of deformation, temperature, molecular structure and solvents.

4.1. Dynamic Measurements
The majority of published experimental data on block copolymers originate in dynamic measurements.

4.1.1. Influence of Frequency
Arnold and Meier[3] investigated a number of S–B–S samples with end-block sizes in the range 10 000–14 000 g mol^{-1} and midblock sizes in the

range 50 000–70 000 g mol^{-1}. The dynamic viscosity data exhibited two 'regions', a high frequency region where the data corresponded to typical behaviour for monomolecular melts and a low frequency region where Newtonian behaviour could not be established, the viscosity increasing continuously with decreasing frequency. The authors formulated the following hypothesis in order to explain the results: With reference to Fig. 1 an ABA block copolymer can exist in three distinct states in a flow field. At rest or at very low rates of deformation a state (A) with an intact three-dimensional network structure exists. The morphology depends on the volume fraction of A groups and two examples are shown in Fig. 1(a) and (b). At intermediate rates of deformation a flow structure (B) consisting essentially of free flowing species of single or multiple domains can exist (Fig. 1(c)).

At high rates of deformation a system (C) where the single polymer molecules constitute the flow units is created during flow. Arnold and Meier found that the transition from microphase (B) to (thermoplastic) monomolecular melt (C) behaviour occurs at a constant stress (τ_{cr}) and is essentially independent of temperature. The critical stress was found to

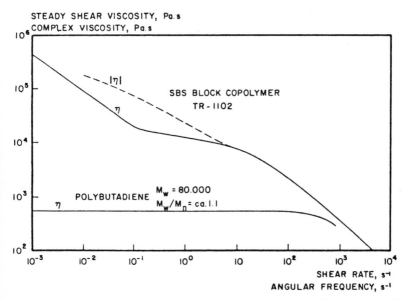

FIG. 4. Viscosity data of Cariflex TR-1102 and polybutadiene at 150°C. Reproduced from Ref. 22 by courtesy of the publisher, the Int. Union of Pure and Applied Chemistry.

be between 4 and $7 \times 10^5 \, \text{dyn cm}^{-2}$ with a best estimate close to $5 \times 10^5 \, \text{dyn cm}^{-2}$.

Qualitatively, these results are in agreement with dynamic data reported in the majority of other papers.[22,25-39,102] This statement does not include the interpretation of the transition B→C mentioned above.

Complementary observations for an S–B–S 11 000–56 000–11 000 show that the ratio between complex viscosity data and steady shear viscosity at identical values of shear rate, $\dot{\gamma}$, and frequency, ω, is much higher than 1 for $\dot{\gamma} < 1 \, \text{s}^{-1}$.[22] Such observations are generally observed for different linear and branched S–B–S samples.[25] Typical data are shown on Fig. 4.

4.1.2. Influence of Temperature

Arnold and Meier show that a double logarithmic plot of reduced viscosity η'/a_T against reduced frequency $a_T \omega$ (where a_T is a shift factor and ω is the frequency) constitutes a reasonable master curve for a given S–B–S block copolymer. The slope in the master curve at low deformation rates varied between -0.36 and -0.66 depending on the styrene/butadiene ratio.

The temperature dependence of the shift factor, a_T, was found to fall on one of two lines when plotted according to the Arrhenius relationship:

$$\ln (a_T) = A + \frac{E}{RT} \qquad (4.1)$$

Arnold and Meier's data are shown on Fig. 5.

Activation energies of $38 \, \text{kcal mol}^{-1}$ and $19 \, \text{kcal mol}^{-1}$ were found for samples with a polystyrene content larger than $35 \, \text{wt}\%$ and less than $31 \, \text{wt}\%$, respectively.

Chung and Gale[26] find that the energy of activation defined by eqn (4.1) changes with temperature for an S–B–S sample with block lengths 7000–43 000–7000. $E = 23 \, \text{kcal mol}^{-1}$ for temperatures less than $150°C$.

4.1.3. Influence of Molecular Structure

Many authors found that ABA polymers have considerably higher viscosities[22] or, expressed in another way, much longer maximum relaxation times for the transition from rubbery to flow behaviour than the corresponding homopolymers with the same molecular weights.

This result is confirmed by Futamura and Meinecke,[27,28] particularly for many different S–X–S polymers where S is styrene and X is varied. They conclude that the maximum relaxation times increase with increas-

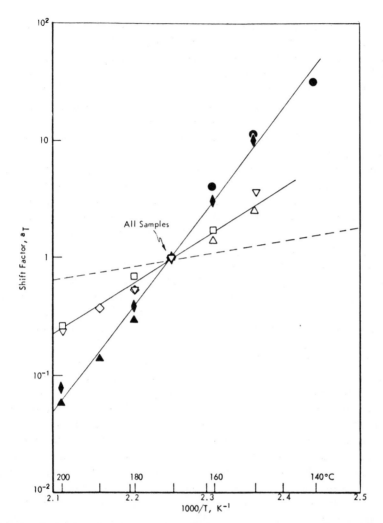

FIG. 5.   Arrhenius plot of a_T: ●, S–B–S 14–50; □, S–B–S 14–60; ◇, S–B–S 14–70; ▲, S–B–S 22–50; ▽, S–B–S 10–50; ◆, MDPS/97; ----, polybutadiene. Reproduced from Ref. 3 by courtesy of the publisher, John Wiley and Sons, Inc.

ing solubility parameter difference between the styrene block and the centre block but are not significantly influenced by the glass transition temperature or the molecular weight between entanglements of the centre block.

The conclusion reached by Futamura and Meinecke is in excellent agreement with the work of Matzner and co-workers[29,30] on organo-siloxane block copolymers. In fact their conclusion was that the differential solubility parameter $|\delta_A - \delta_B|$ (δ measured in $(cal\,cm^{-3})^{1/2}$) should be less than one in order to obtain good melt processability. Furthermore, these authors reached the important conclusion that optimum process-ability and two-phase properties will be obtained in block and graft co-polymers if the differential solubility parameter is ~zero, and if at least one of the segments is crystallizable.

An important contribution which increases the possibility of the interpretation of structure transitions in block copolymers comprises the relatively large number of papers which deal with investigations of dynamic data as a function of temperature on block copolymers showing a thermodynamic transition from non-compatible at low temperature to compatible at high temperatures.[26,31-34]

Kraus et al.[25] observed Newtonian behaviour at 160°C for an S–B–S sample with low total molecular weight (54 000, 30% PS); however, no explanation was given of this observation. Chung and Gale[26] were the first researchers to perform a systematic investigation on an S–B–S sample (S:7000, B:43 000) showing a structure transition by increasing the temperature.

In fact, Leary and Williams[40] defined such a thermodynamic critical equilibrium temperature T_S for ABA copolymers, above the highest glass transition temperature of any of the copolymer phases, at which the free energy of mixing is zero. Consequently, below T_S a multiphase structure is stable. Above T_S a statistical mixing — a homogeneous state — is stable. Here, phase miscibility is obtained by decreasing the molecular weight of the blocks and by increasing the temperature. The treatments of Krause,[41] Meier[42] (for diblock copolymers) and Helfand[43] could equally well be used to explain the transition to homogeneous state.

Widmaier and Meyer[34] found T_S for a styrene/isoprene/styrene, S–I–S, block copolymer and documented the transition to homogeneous phase with melt rheological, electron microscopy, X-ray and light diffraction measurements. Chung et al.[33] used electron microscopy as well as rheological measurements to verify the transition at T_S.

Thus the occurrence of a thermodynamical transition at a critical temperature T_S rests on a sound theoretical and experimental basis.

An instructive example for an S–B–S sample is shown in Fig. 6 from Gouinlock and Porter[31] and for S–I–S in Fig. 7 from Widmaier and

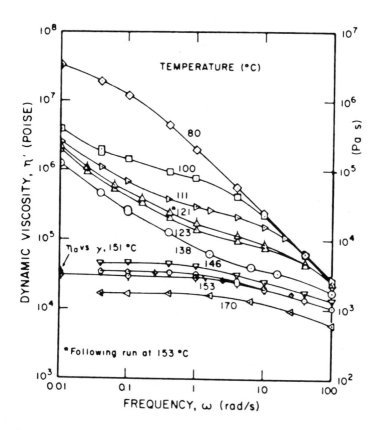

FIG. 6. Dynamic viscosity vs. frequency for an S–B–S sample. Steady-state data, apparent viscosity vs. shear rate (s^{-1}), indicated. Upper 153°C curve is for a toluene-cast specimen, lower curve is for a cyclohexane-cast specimen. Reproduced from Ref. 31 by courtesy of the publisher, the Society of Plastics Engineers, Inc.

Meyer.[34] The value where both the dynamic elastic modulus and the viscosity form one curve corresponds to $\tau_{cr} = \eta'\omega = 5{\cdot}5 \times 10^5$ dyn cm^{-2} for S–B–S, in excellent agreement with the data of Arnold and Meier.[3]

4.1.4. Influence of Solvents
A number of papers are concerned with investigations of solutions of block copolymers[44-51] (dilute solution properties not included).

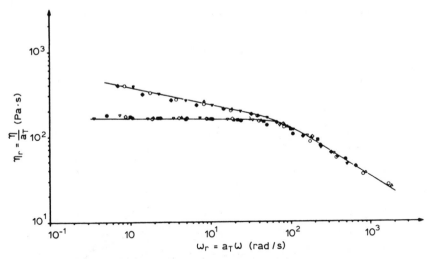

FIG. 7.   Reduced viscosity vs. reduced frequency for an S–I–S sample. Shift
factor a_T. $T(°C)$, a_T: △, 170, 0·03; ▼, 185, 0·10; ○, 200, 0·29; ◆, 215, 0·76; ●, 225,
1; ▽, 230, 1·48; ■, 235, 2·26; ◇, 240, 2·64; ▲, 250, 5·20. Reproduced from Ref. 34
by courtesy of the publisher, John Wiley and Sons, Inc.

The influence of low molecular mass solvents on the properties of
solutions of block copolymers depends strongly on the interaction of the
solvent molecules with each of the comonomers (components) of the
block copolymer. If the solvent is common to both components the main
effect on the properties seems to originate in a depression of the
thermodynamic critical equilibrium temperature T_S. Pico and Williams[49]
have extended their theory for the prediction of T_S to encompass
plasticized triblock copolymers. The major result is that T_S is predicted
to be depressed when the solvent fraction, ϕ_S, is increased. The depression
is maximal for solvents which have a solubility parameter, δ_S, equidistant
between the solubility parameters, δ_A and δ_B, for each of the components
of the block copolymer, but becomes rapidly less when δ_S is outside the
δ_A–δ_B range. For block copolymers plasticized with common solvents
investigated at temperatures below T_S, a behaviour equivalent with that
reported for undiluted block copolymers is expected and observed (for
concentrated S–B–S melts) as reported by Pico and Williams.[50] For
plasticized samples investigated above T_S, the behaviour of the single-
phase solution formed is expected to be quite analogous with the
behaviour of monomolecular polymer solutions and this behaviour is

observed.[50] However, if selective solvents are used, e.g. a solvent for one component but a non-solvent for the other component, various anomalies may appear.[51]

4.2. Simple Shear Flow

Transitions observed in steady shear flows such as a transition between microphase and thermoplastic (monomolecular) behaviour at a critical stress are more directly understandable than a transition during dynamic measurements. However, the similarity between dynamic viscosity and steady-state viscosity–shear rate curves observed for monomolecular melts (e.g. the Cox–Merz rules[52]) and block copolymers indicates that the results from dynamic measurements may be interpreted in the same way as steady shear flow results. This could tentatively be explained by 'the principle of minimum viscous dissipation' as for stratified two-phase flow and used by Lyngaae-Jørgensen[14] in order to rationalize the similarity of shear viscosity and dynamic viscosity data.

Data measured in cone and plate geometry or rotary viscometers[22,24,26,35,53-57,74] and data measured in capillary or slit flow[22,24,25,35,54,58-70] have been extensively used in investigations of block copolymers.

4.2.1. Influence of Shear Rate

One of the first systematic investigations on flow properties of S–B–S block copolymers at high temperatures, in both cone and plate and capillary flow, was reported by Holden et al.[54] Holden et al. observed that one of the most conspicuous differences between the rheological properties of block copolymers and homopolymers is that the block copolymers do not exhibit Newtonian behaviour at low shear rates (see Fig. 8).

Furthermore, it was found that materials with a styrene content in the range 39–65% showed two distinct viscosity shear rate relationships (see Fig. 9). The authors interpret this fact as being a result of phase inversion. It is observed that transitions from one state to the other occur at about the same shear stress, which Holden et al. find to be approximately 10^6 dyn cm^{-2}, independent of temperature.

The IUPAC group report that low shear rates cone and plate measurements indicate that shear stress–time curves have very different forms depending on the shear rate (see Fig. 10). In many cases no well-defined steady value for the shear stress is found. Consequently, the calculation of viscosity becomes rather arbitrary and depends on how shear stress is evaluated. The scatter between the values from different

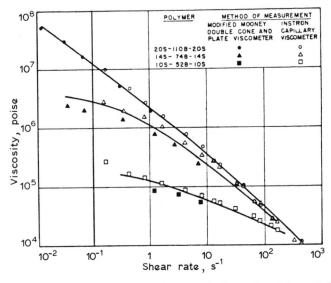

FIG. 8.   Viscosities of S–B–S polymers at 175°C. Reproduced from Ref. 54 by courtesy of the publisher, John Wiley and Sons, Inc.

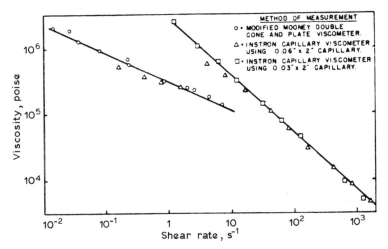

FIG. 9.   Viscosity of a polymer, with a styrene content of 39%, at 150°C. Reproduced from Ref. 54 by courtesy of the publisher, John Wiley and Sons, Inc.

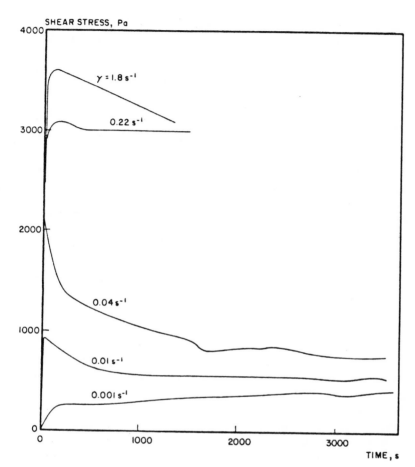

FIG. 10.   Typical shear stress–time traces at 150°C. Reproduced from Ref. 22 by courtesy of the publisher, the Int. Union of Pure and Applied Chemistry.

laboratories was found to be considerable. It was documented that prehistory played a very significant role. At higher shear rates (higher shear stresses) the viscosity was much more reproducible and in agreement with capillary data.

This observation is in accord with the work of Vinogradov et al.[24] who, for the same S–B–S sample investigated by the IUPAC group, report that the material in the region of low shear stresses exhibits sharply pronounced thioxotrophy. Vinogradov et al. found that in the region 3×10^4 to 3×10^5 dyn cm^{-2} nearly Newtonian flow with viscosity, η_0, of the material is observed in capillary flow (150°C). This observation is in reasonable accordance with the IUPAC data as shown on Fig. 4. Vinogradov et al.[24] found that capillary flow data follow a time–temperature position principle of the type $\log \eta/\eta_0$ against $\log (\eta_0 \dot\gamma)$.

Vinogradov et al. report that 'spurt behaviour' ($\sim$melt fracture) at the capillary wall is observed at $\tau_{cr} = 3 \cdot 5 \times 10^6$ dyn cm^{-2} for the sample investigated by them and the IUPAC group. This phenomenon will be discussed in more detail in Section 6.1.

Ghijsels and Raadsen[22] interpret the IUPAC result as being caused by structural changes taking place in the melt (breaking of network structures) at low shear stresses. For shear stresses above $\sim 3 \times 10^4$ dyn cm^{-2} star-shaped branched flow units are taken as responsible for the flow behaviour. This star structure may eventually be broken down at high stresses.

Williams and co-workers[99,100] have shown that S–B–S triblock copolymers exhibit yield stress behaviour and have developed a special controlled-stress, parallel-plate rheometer for measurement of yield stresses.

4.2.2. Influence of Temperature

The IUPAC group reports that the viscosity at a shear stress 10^4 Pa of the S–B–S block copolymer (TR-1102) does not follow an Arrhenius type of temperature dependence over wide intervals in temperature. The activation energy increases with decreasing temperature; E is about 15 kcal mol^{-1} at temperatures between 170 and 190°C and about 30 kcal mol^{-1} at temperatures between 110 and 130°C. Holden et al.[54] also report energies of activation (varying with temperature) at $\tau = 3 \times 10^4$ Pa for different S–B–S samples.

4.2.3. Influence of Structure

Holden et al.[54] found a much higher exponent than $3 \cdot 5$ in the relation $\eta(\dot\gamma = 10^{-1} \text{ s}^{-1}) = K M_w^a$ observed for S–B–S samples with 27% styrene content and total molecular weights ranging from 73 000 to 150 000 (see Fig. 8).

The viscosities at $\tau = 2 \times 10^4$ Pa and $T = 75°C$ of a series of six S–B–S samples, with total molecular weight approximately 80 000 and styrene content varied from 13% to 80% by weight, were compared with the viscosities of PB and PS with the same molecular weight at the same conditions. Holden et al. found that the viscosity at $\tau = 2 \times 10^4$ Pa (low shear rates) goes through a pronounced maximum with increasing styrene content and the viscosities for the S–B–S samples were higher than the viscosity of either homopolymer.

Kraus et al.[25] investigated different samples of B–S–B and S–B–S and three- and four-branched samples. The authors show that S–B–S samples at $\dot{\gamma} = 1\,\mathrm{s}^{-1}$ and $\dot{\gamma} = 10\,\mathrm{s}^{-1}$ and $T = 130°\mathrm{C}$ and $T = 160°\mathrm{C}$ have higher viscosities than B–S–B samples and that the viscosity of the B–S–B samples is determined by the butadiene block lengths.

Leblanc[59,60] reports data measured with a slit die rheometer on a star-shaped butadiene/styrene block copolymer (40% styrene by weight, $\bar{M}_w = 153\,000$ $\bar{M}_w/\bar{M}_n = 1·24$). Above a critical shear stress, τ_{cr}, equal to $5·5 \times 10^5\,\mathrm{dyn\,cm}^{-2}$, a power law behaviour with the same flow index is observed at all temperatures. Below this critical stress a marked curvature and discrepancy between data from slit dies of different depths are observed. For data above τ_{cr} a flow activation energy at constant stress is $19·2\,\mathrm{kcal\,mol}^{-1}$.

Very few systematic investigations are reported on diblock copolymers.[35,53,65] Nguen Vin Chii et al.[65] compare the flow behaviour of a diblock copolymer of polyisoprene and polybutadiene PI–PB with 30% PB and a total molecular weight of $3·5 \times 10^5$ in the temperature range 25–90°C. The flow behaviour of the block copolymer is compared with the behaviour of homopolymers of PB and PI and blends of these homopolymers. The PI–PB block copolymer exhibited a Newtonian behaviour at low shear rate. It is reported that the zero shear viscosity of the block copolymer is approximately twice the zero shear viscosity of either homopolymer for samples with approximately the same total molecular weight by weight. The mixing rule

$$\eta_0 = (w_1 \eta_{0,1}^{1/\alpha} + w_2 \eta_{0,2}^{1/\alpha})^\alpha \qquad (4.2)$$

is reported to be followed for blends of samples of the same type of homopolymer;[85] w_1 and w_2 are weight fractions of components with viscosities $\eta_{0,1}$ and $\eta_{0,2}$ and α is the exponent in the molecular weight dependence of η_0. Equation (4.2) did not apply for the investigated blends of PI and PB nor for the diblock copolymer. The flow energy of activation for the block copolymer is found to be between the energy of activation for the homopolymers.

Lyngaae-Jørgensen et al.[53] prepared a diblock copolymer of polystyrene and poly(methyl methacrylate) with 25% PMMA and a total molecular weight by weight of 91 000. Transitions were observed in the flow curves, in qualitative agreement with those reported for S–B–S block copolymers, but the transitions observed at different temperatures did not take place at constant shear stress.

Data for multi-block copolymers of type A_nB_n are rather few in number.[35,61] Anomalous flow phenomena as non-Newtonian behaviour at low shear rates and thixotropy, qualitatively as for S–B–S materials, have been observed for block copolymers of ethylene and propylene[35] but at very low shear rates (shear stresses).

4.2.4. Influence of Solvents

A number of papers discuss the properties of solutions of block copolymers during shear flow.[56,57,71,72,74,76,77,101,108] As mentioned in Section 4.1.4 the polymer solvent interaction is important. The rheological behaviour should be analogous to the behaviour of monomolecular polymer solutions at temperatures above the thermodynamic critical temperature, T_S.

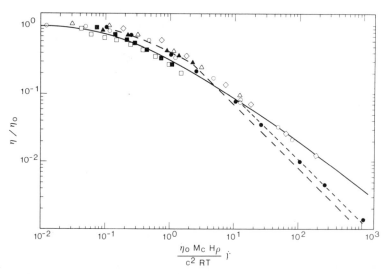

FIG. 11.   Double logarithmic plot of reduced viscosity, η/η_0 against reduced shear rate. ●, S–B–S, 28 wt%S, $\bar{M}_n = 78\,000$, $H = 1\cdot2$, $T = 150°$C (Ref. 22); △, S–B–S, 28 wt%S, $\bar{M}_n = 78\,000$, $H = 1\cdot2$, $T = 150°$C (Ref. 24); ○, S–B–S, 28 wt%S, $\bar{M}_n = 78\,000$, $H = 1\cdot2$, $T = 170°$C (Ref. 22); ■, S–B–S, 40 wt%S, $\bar{M}_w = 129\,000$, $H = 1\cdot2$, $T = 200°$C (Ref. 53); ◇, S–B–S in tetralin, 31 wt%S, $\bar{M}_n = 105\,600$, $H \approx 1$, $T = 35\cdot5°$C, $c = 0\cdot306\,\mathrm{g\,cm}^{-2}$ (Ref. 71); □, PS–PMMA, 25 wt%PMMA, $\bar{M}_w = 91\,000$, $H = 1\cdot82$, $T = 160°$C (Ref. 53); ▲, PB–PI, 30 wt%PB, $M \simeq 3\cdot5 \times 10^5$, $T = 25°$C (Ref. 65).

In fact, solutions of block copolymers in common solvents do exhibit this behaviour.[71,75] Paul et al.[71] showed that viscosity data for solutions of an S–B–S block copolymer in five different solvents could be super-imposed to form a master curve with the same curve form as Greaseley's master curve for monodisperse monomolecular melts, when plotted as reduced viscosity η/η_0 against $\lambda_0\dot{\gamma}$, where λ_0 a characteristic time was found by a statistical curve fit program. One of the data sets of Paul et al. is plotted in Fig. 11 for comparison with the master curve treated in Section 6.1. The fitted characteristic times, λ_0, of Paul et al. are roughly equal to $(3\eta_0 M_c H\rho)/(c^2 RT)$, which was found by Lyngaae-Jørgensen[14] to be in accord with stress relaxation data.

4.3. Simple Elongational Flow and Supplementary Measurements[22,103]

The IUPAC group have reported that the elongational viscosity of the IUPAC S–B–S sample is tension thinning in the stress region 10^5–10^6 Pa at 150°C. Slow stretching of an extrudate at 150°C leads to necking. Die swell measurements in this sample show a very low value at shear stresses less than approximately 10^5 Pa but increase sharply with increasing stress above 10^5 Pa. Structural changes are investigated by flow birefringence in a slit capillary and indicate a structure transition at approximately 4×10^5 Pa.

Furthermore, shear stress relaxation data and birefringence measurements show varying degrees of incomplete relaxation.[22,23] Steady normal stress data could only be obtained for $\dot{\gamma} > 1\ \mathrm{s}^{-1}$.

5. HYPOTHESES CONCERNING FLOW MECHANISMS AND STRUCTURE TRANSITIONS

Earlier sections demonstrate that the rheological behaviour of AB, ABA and $A_n B_n$ block copolymers with non-compatible blocks is considerably different from the behaviour of that of homopolymers as well as random copolymers. The block copolymers are reported to have higher melt viscosities than random copolymers with the same composition and weight average molecular weight, at least at low shear stresses. The viscosity of multiphase block copolymers does not approach a constant viscosity at low shear rates. Furthermore, abrupt changes of slopes in flow curves have been observed by many research groups.

There is certainly no general consensus concerning the flow units and flow mechanisms. However, it seems that almost all authors attribute the complex melt flow behaviour to the existence of the two-phase structure and breakdown of this structure during flow. Such structure breakdown is, as noted by many authors in the field, not exclusive for block copolymers but is observed for filled materials,[2] ethylene/propylene copolymers,[35] ABS copolymers and PVC materials.[78] Structure transitions are also in these cases the probable cause for the observed flow behaviour. To summarize the main hypothesis regarding the structure–flow unit problem, one can state that the following has been formulated:

Holden et al.[54] rationalized the observation that the measurements were dependent on prehistory, i.e. on the measuring instrument, as a consequence of the existence of two different states in the melt. The observed change in flow behaviour at approximately 10^5 Pa was attributed to a phase transition. Arnold and Meier[3] formulated a hypothesis for the structure changes of an ABA block copolymer in the observed viscosity–frequency shear rate dependency. At rest or at very low deformation rates the material behaves as a three-dimensional network by virtue of the two-phase structure. At intermediate deformation rates this structure will gradually be disrupted, forming flowing domains and aggregates of many molecules. The system will behave as large star-shaped aggregates. Finally, at high deformation rates the system will break down to a monomolecular melt state. Arnold and Meier estimated that the transition to a monomolecular melt state takes place at an approximately constant stress $\sim 5 \times 10^4$ Pa.

A systematic approach towards a quantitative model has been formulated by Henderson and Williams.[79] This approach is capable of 'explaining' in a qualitative way many of the experimental observations. A two-phase structure is considered with pure A domains in a matrix of pure B separated by interphase regions of mixed composition, with a composition profile over the interphase thickness, ΔR. In shear flow (or other flows) three events are considered: pulling out of an A block from an A domain, migration through a B matrix, and rejoining another A domain. The forces acting on a polymer molecule are frictional forces, F^f, which consists of four terms: drag on B segments moving through B matrix plus drag on A segments moving through the B matrix plus interphase drag on each segment type. The total withdrawal force, F, on the chain consists of two contributions, namely the frictional forces, F^f, plus an interphase thermodynamic contribution, F^i. The chemical potential of each component varies through the interphase, e.g. for component A: $\mu_A(r)$, where r varies from zero to the interphase thickness, ΔR. The chemical potential profile gives rise to an interphase barrier force,

$F^i_A(r) = -\nabla\mu_A$, resisting the mixing processes. Since μ_A depends on the interaction of A and B segments the barrier force F^i and $\eta(\dot{\gamma})$ must be sensitive to $\delta_A - \delta_B$ as observed in the literature. Establishment of an equilibrium state is predicted to be time dependent since it involves slow diffusional processes through interphases, thus explaining the thixotropy observed for block copolymers. Furthermore, the equilibrium state (steady state) itself will be highly shear rate dependent, one reason being the mechanism cited above depends on the time permitted for re-entry of A blocks from the B matrix to A domains. Penetration time is decreased when $\dot{\gamma}$ is increased.

Finally, Lyngaae-Jørgensen et al.[53] have formulated a criterion for a transition from a two-phase state to a monomolecular melt state in block copolymers. This approach may be used to roughly evaluate the influence of structure variables on processing properties; a form of the original modified approach is presented in the appendix.

The final expression in the limiting case for average number degree of polymerization is as follows: $\bar{X}_n \to \infty$ and the interaction parameter roughly estimated from

$$\chi = \frac{\bar{V}}{RT}(\delta_A - \delta_B)^2 \tag{5.1}$$

predicts that a transition to a monomolecular melt state takes place at a critical shear stress τ_{cr} given by the equation:

$$\tau_{cr} = \left(k \cdot \rho \frac{V_A}{\bar{V}}\frac{RT}{M_c}\left(1 - \frac{2}{z}\right)^{1/2}\frac{|\delta_A - \delta_B|}{H}\right) \tag{5.2}$$

where k and z are constants (given in the appendix), V_A and $\bar{V}$ and the molar volume of monomer A and the average molar volume of repetition units, R is the gas constant, T is absolute temperature, M_c is twice the molecular weight between entanglements, H is $\bar{M}_w/\bar{M}_n$, and δ_A and δ_B are the solubility parameters of monomer A and monomer B, respectively.

This expression can only be a rough first approximation because, for example, eqn (5.1) is highly questionable.

Since this chapter was written Ekong and Jayaraman[109] have published a rheological equation based on a kinetic network model for the rheology of block copolymer melts with spherical domains and filled polymer composites.

6. DATA STRUCTURE ANALYSIS USING MASTER CURVE DATA REPRESENTATION FOR VISCOSITY DATA

As reviewed in Section 4 many research groups have shown that superposition of viscosity–shear rate data measured by different temperatures to one single master curve is possible.[3,24,31,71] For linear homopolymers in the monomolecular melt state the relations

$$\frac{\eta}{\eta_0} = F\left(\frac{\eta_0 M_c H\rho}{c^2 RT} \cdot \dot{\gamma}\right) \qquad (6.1)$$

and

$$\frac{\eta'}{\eta'_0} \quad F\left(2\frac{\eta_0 M_c H\rho}{c^2 RT}\omega\right) \qquad (6.2)$$

constitute a reasonable approximation with the same functional form, at least at large shear rates for a number of polymers, including PS, PMMA, poly(vinyl acetate), polyisoprene, polybutadiene, polyoxymethylene, Nylon 6 and Nylon 66 and polyacrylamide.[14] Thus, above τ_{cr}, this master curve should be applicable for, for example, S–B–S, S–I–S and PB–PI, PS–PMMA block copolymers. However, in order to analyse a set of data using eqn (6.1) and/or eqn (6.2) an estimate of η_0 as it would be for a monomolecular melt is necessary. This requirement make the approach questionable and in some cases inapplicable. Unambiguous direct structure observations (see Section 4) are not accessible at present.

However, lack of agreement with eqn (6.1) and eqn (6.2) can be used to reject the supposition that a monomolecular melt state may exist in the melt. Thus, rheological data analysed as described above may be used to possibly reject a hypothesis concerning a transition to a monomolecular state.

For samples showing a Newtonian range, a zero shear viscosity is defined as the Newtonian viscosity. Ferry's equation

$$\frac{1}{\eta} = \frac{1}{\eta_0^*} + b\tau \text{ for } \tau > \tau_{cr}$$

is used to define a zero shear viscosity η_0^* for the assumed monomolecular melt state. For data where τ_{cr} is close to the Newtonian range this empirical procedure should give reasonable estimates.

Figure 11 shows a compilation of data from simple shear flows plotted in accordance with eqn (6.1). The full curve is the master curve for monomolecular melts of linear homopolymers. Data for PB–PI, S–B–S and PMMA–PS block copolymers are plotted for $\tau > \tau_{cr}$. Two features are apparent from the plot: (a) the data scatter roughly around the master curve, and (b) the data reveals what Vinogradov has called the spurt phenomenon at high shear rates.

Thus, based on a comparison between measured data and the master curve for homopolymers, one cannot reject a hypothesis which assumes that the data represent rheological properties of a monomolecular melt.

6.1. Fracture Phenomena

At high deformation rates (or high stresses) a transition from the fluid to a forced high elastic (rubbery) state (FHES) may take place in concentrated polymer melts. This concept is due to Vinogradov and co-workers.[82–84] The lowest critical stress corresponding to a transition to FHES which is a constant for each homologous polymer series, is equal to the maximum value of the loss modulus, determined from dynamic testing. A transition to FHES inevitably leads to fracture of the polymer and indicates the maximum production rate obtainable with conventional processing methods without melt fracture. Melt fracture may occur at the inlet to the capillary in a capillary rheometer or in the capillary at the wall and will (normally) result in distorted extrudates.

In capillary flow a transition to FHES for the material in the capillary is revealed by an abrupt change of slope to zero slope in the flow curves ($\log \tau$ against $\log \dot{\gamma}$). In a double logarithmic delineation of viscosity against shear rate the spurt is depicted as a straight line with slope of -1. Vinogradov found a critical stress equal to $3.5 \times 10^6 \, \mathrm{dyn\,cm^{-2}}$ for the S–B–S sample studied by the IUPAC group.[24] Vinogradov's spurt curves are shown in Fig. 11 as dashed lines.

Thus, from a processing point of view, this spurt phenomenon, which may be observed either in a delineation of shear flow data or in a delineation of dynamic data, is important, one reason being that it is the upper limits for the area where processing of items with smooth surfaces in extrusion may be produced.

The melt fracture phenomenon is often observed at smaller stresses than those corresponding to spurt fracture where the fracture takes place near the capillary wall. This may be caused by fracture in the entrance zone to the capillary and triggered by the predominantly elongational flow which take place in tapered inlet zones.[14]

The difference between the critical stress, τ_{cr}, for a transition to a monomolecular melt state and the spurt stress in the monomolecular melt state are rather small for S–B–S samples. For high molecular weight samples a value of $\tau_{cr} = 2 \cdot 9 \times 10^6$ dyn cm^{-2} is predicted from eqn (5.2). This means that high molecular weight samples pass directly from an aggregate state to the spurt range. This prediction seems to be substantiated by the data of Holden et al.[54] and Arnold and Meier.[3]

The lowest possible viscosities and consequently the easiest processing of a block copolymer are obtained under conditions where the material is close to, or in, a monomolecular melt state. Thus, fast shear rates (production rates) are advantageous in order to obtain structure breakdown. Transition into a forced high elastic state will usually limit the useful production rate range. The slip flow phenomenon may perhaps be used to give very fast mould-filling operations in injection moulding. Such applications do, however, presuppose very careful mould design in order to avoid jetting and weld lines.

Only for samples with relatively low molecular weight near the transition temperature will a transition to a monomolecular melt state be possible for S–B–S samples. In fact S–B–S samples seem to constitute a marginal position in this respect. However, the S–B–S sample studied by the IUPAC Working Party seems nevertheless to have been in a monomolecular melt state at least at temperatures higher than $\sim 150°C$.

This conclusion may be drawn by comparison of zero shear viscosity data measured above the temperature where one-phase behaviour is observed (T_S) with the zero shear viscosities estimated from Ferry's equation.

In Table 1 is shown a compilation of data from Gouinlock and Porter[31] with η_0^* data estimated from Ghijsel's paper (the IUPAC report).[22]

TABLE 1

	Styrene content (wt%)	Molecular weight	$\eta_0^*(T = 150°C)$ (P)	$\eta_0^*(T = 170°C)$ (P)
Sample 1	24·8	57 000	$3 \cdot 7 \times 10^4$	$1 \cdot 7 \times 10^4$
Sample 2	28	78 000	1×10^5	$4 \cdot 9 \times 10^4$

$\dfrac{\eta_2(150°C)}{\eta_1(150°C)}$	$\dfrac{\eta_2(170°C)}{\eta_1(170°C)}$	$\left(\dfrac{M_2}{M_1}\right)^{3,4}$
2·70	2·88	2·74

These data, together with the fact that the viscosity data scatter around the master curve for homopolymers (Fig. 11), may be taken as an indication that the sample is in a monomolecular melt state. For the IUPAC sample eqn (A.9) gives the estimate $\tau_{cr} = 8.5 \times 10^5$ dyn cm^{-2}, meaning that a monomolecular melt state should be expected in a stress interval around 10^6 dyn cm^{-2}.

A knowledge of how a monomolecular melt would behave under given experimental conditions can often facilitate the analysis of melt flow data for melts with complex melt structures. However, such a prediction presupposes a knowledge of mixing rules, e.g. for prediction of the zero shear viscosity of a block copolymer from a knowledge of its constituents.

Surprisingly, the semi-empirical mixing rule found by Friedman and Porter[85] for mixtures of samples of the same homopolymer:

$$\eta_0^{1/a} = w_1 \eta_{0,1}^{1/a} + w_2 \eta_{0,2}^{1/a} \qquad (6.3)$$

seems to work well for data for a block copolymer (PMMA–PS). The data are given in Table 2.

TABLE 2

COMPARISON OF ZERO SHEAR VISCOSITY DATA CALCULATED ACCORDING TO EQN (6.3) AND EXPERIMENTAL DATA FOR DIBLOCK COPOLYMER PMMA–PS (25/75) WITH $\bar{M}_w = 91\,000$

Temperature ($^\circ$C)	$\eta_0^{PMMA} \times 10^{-3}$ (Pa s)	$\eta_0^{PS} \times 10^{-2}$ (Pa s)	$\eta_{0,m}^{calc} \times 10^{-2}$ (Pa s)	$\eta_{0,m}^{exp} \times 10^{-2}$ (Pa s)
170	486·3	58·92	178	385
180	230·4	27·92	85·3	83·3
190	112·7	13·66	42·1	41·7
200	56·86	$\eta_0^{ref} =$ 6·89	21·5	21·8
210	$\eta_0^{ref} =$ 29·5	3·58	11·3	11·3
220	15·72	1·91	6·08	6·0
230	8·59	1·04	3·37	3·3
240	4·80	0·582	1·91	1·9
250	2·72	0·330	1·01	1·1

$$(\eta_{0,m})^{1/a_m} = w_1 \eta_{0,1}^{1/a_1} + w_2 \eta_{0,2}^{1/a_2} \qquad (6.4)$$
(Ref. 85)

and

$$a_m = w_1 a_1 + w_2 a_2$$

$$\log \eta_0 = 3.93 \log \bar{M}_w - 14.02 \qquad (6.5)$$
(Ref. 86)

for PMMA at 210°C.

$$\log \eta_0 = 3 \cdot 17 \log \bar{M}_w - 11 \cdot 97 \qquad (6.6)$$
(Ref. 87)

for PS at 200°C.

$a_1 = 3 \cdot 93$, $a_2 = 3 \cdot 17$, $a_m = 3 \cdot 36$, where w_1 and w_2 are volume fractions of polymer 1 and polymer 2, and $\eta_{0,1}$ and $\eta_{0,2}$ are the zero shear viscosities of polymer 1 and polymer 2, respectively. Energy of activation = 30 kcal mol^{-1} for both PS and PMMA.

Such calculations are, however, enormously sensitive to systematic errors in the zero shear viscosity–molecular weight relations used in the calculations. In particular, systematic deviations between estimates of the average molecular weight by weight $\bar{M}_w$ will give very large differences between calculated zero shear viscosities because of the $\sim 3 \cdot 4$ exponential dependence.

Such systematic deviations between different investigators are often observed[88] and great caution must therefore be exercised when comparing calculated and measured viscosity data.

As an example, measurements of corresponding values of zero shear viscosities and weight average molecular weights (GPC) on polybutadienes with given cis-1,4; trans-1,4; and trans-1,2 content at this institute leads to $\eta_0 - \bar{M}_w$ relations which give much larger estimates of η_0 than the data measured by Kraus and Gruver.[89,90] This deviation may be caused by systematic deviations in the $\bar{M}_w$ measurements as a result of different calibration procedures, measuring technique, etc. The master curve relations used above seem to be less vulnerable to this kind of uncertainty if all data used are based on experimental results from the same source.

6.2. Discussion of Structure Transitions

It may be argued that optimal processing properties are obtained for melts which are in or near a monomolecular melt state at processing conditions.

The hypothesis presented in Section 5 clearly postulates that a monomolecular melt state is established above a critical shear stress, τ_{cr}. The expression for τ_{cr} given in eqn (A.9) is in qualitative agreement with many of the experimental observations as discussed in Section 5. Nevertheless, the experimental evidence is insufficient for a rigid test of the predictive validity of the expression. However, the expression seems capable of allowing qualitative conclusions (trends) concerning the influence of structure variables, e.g. $|\delta_A - \delta_B|$ on rheological properties.

The validity of the basic postulate that a monomolecular melt state can be obtained for block copolymers with non-compatible blocks (at $\tau = 0$), rests on the following experimental observations. Melts of block copolymers give rise to flow curves with abrupt changes of slopes. Dynamic and steady shear flow data merge at high frequency/ high shear rate. Above a critical shear stress some block copolymers show the same structure viscosity relationships as expected for linear homopolymer melts in steady simple shear flow and dynamic shear flow (Figs 11 and 12). The morphological structure of rapidly cooled samples of a PS–PMMA diblock copolymer changes to a nearly homogeneous state. The estimated zero shear viscosity of this block copolymer follows a simple mixing rule as shown in Table 2. The zero shear viscosity estimated for S–B–S samples above the transition temperature where a homogeneous melt is formed, T_S, is in reasonable agreement with the estimated zero shear viscosity for an S–B–S sample with higher molecu-

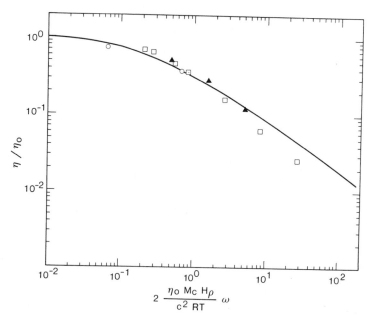

FIG. 12. Double logarithmic plot of reduced viscosity, η/η_0 against reduced frequency. ▲, S–B–S, 28 wt%S, $\bar{M}_n = 78\,000$, $H = 1\cdot2$, $T = 150^\circ$C (Ref. 22); □, S–B–S, 28·6 wt%S, $\bar{M}_n = 70\,000$, $T = 170^\circ$C (Ref. 3); ○, S–B–S, 24·6 wt%S, $M = 57\,000$, $T = 170^\circ$C (Ref. 31).

lar weight but under conditions where $T < T_S$ as judged from the Flory–Fox relation $\eta_0 = KM_w^{3\cdot4}$ (Table 1).

The work presented suggests that a monomolecular melt state might be a stable flow state even for systems which form separate phases in the unsheared state. The experimental evidence so far produced is not in contradiction to this prediction, but, until direct unambiguous measurements of structure in a flowing system of non-compatible polymers have shown that it is in fact a one-phase system under flow conditions, the one-phase formation is still only a working hypothesis.

Finally, a few comments on experimental difficulties in connection with direct observations on melts of block copolymers during flow and at high rates of deformation.

One complication which may be exemplified by discussion of capillary measurements is that most of the equipment which can be used at high stresses or high deformation rates have a stress distribution. For flow in a capillary the stress is zero in the centre and maximal at the wall. This fact makes it difficult to interpret birefringence measurements, for example, since the melt structure may change over the radius of the capillary. Another complication may arise from the finite residence time which the capillary instruments, for example, encompass.

7. SUMMARY OF ESSENTIAL FEATURES

The flow behaviour of multiphase block copolymers is indeed complex. As described above, microstructure changes do often take place during flow. At present it is fair to say that the efforts towards a better understanding of the flow behaviour of block copolymer systems have not advanced far. An approach towards a systematic treatment have been sketched[79] but a development towards a valid rheological equation of state seems remote at present. The subject has important technological aspects, e.g. the development of a better understanding of the flow behaviour of mixtures of block copolymers with blends of homopolymers in order to enhance the interfacial adhesion.

However, the limited knowledge which is available, especially from studies in simple flows, allows some guidelines for the development of new products and a better use of the existing ones.

The only way to increase the monomolecular melt range seems to be to change the molecular structure of the polymer molecules.

The conclusions reached by Matzner and co-workers[29,30] that good processability requires $|\delta_A - \delta_B| < 1$, that optimal processing properties should be obtained for $|\delta_A - \delta_B| \simeq 0$ and that at least one component should be crystalline are roughly in accordance with the criterion $\tau_{cr} < \tau_{spurt}$.

Apart from minimizing $|\delta_A - \delta_B|$ an increasing heterogeneity, H, would tend to give better processing properties for constant $\bar{M}_w$ since τ_{cr} decreases (eqn (5.2)). Finally, decreasing molecular weight and possibly increased branching ease processing considerably, as has been known for many years.

Thus in summing up, when developing new block copolymers or choosing block copolymers for given uses, the structure parameters given in Table 3 are most sensitive from a processing point of view.

TABLE 3

Parameter	Parameter change	Processability
Compatibility (first approx.) $\|\delta_A - \delta_B\|$	↓	+
Number average degree of polymerization: $\bar{X}_n$	↓	+
Heterogeneity: H	↑	+
Polymer concentration in solvents for both block components: c	↓	+

The temperature plays a dominant role on domain stability in a temperature range close to the critical transition temperature. At present the temperature superposition principle used by Arnold and Meier[3] seems to be the best data correlation; that is, for a given sample a plot of say η/a_T against $a_T\dot{\gamma}$ or $a_T\omega$, where a_T is a shift factor gives a master curve.

The approach outlined in Section 6 (eqn (6.1) and eqn (6.2)) is only useful for systems where η_0^* can be estimated.

To answer some of the unsolved problems concerning structural changes during flow more systematic investigations are required, involving development of direct structure evaluation techniques.

8. RHEOLOGY IN SIMPLE FLOW FIELDS AND PROCESSING PROPERTIES OF COMMERCIAL BLOCK COPOLYMERS

The treatment in the previous sections has shown that the rheological properties of two-phase block copolymers do involve a gradual structure breakdown with increasing shear rates or shear stresses. It can also be inferred that with all other factors unchanged, high shear rates are advantageous since an almost monomolecular melt state can then be approached depending on variables such as $|\delta_A - \delta_B|$, $\bar{M}_w$, etc. The fabrication operations normally used for processing of thermoplastic materials do normally operate at high shear rates. The different shear rate ranges usually encountered in manufacturing processes are:[91]

$$\text{Compression moulding: } \dot{\gamma} = 1 \quad \text{to } 10 \text{ s}^{-1}$$

$$\text{Calendering: } \dot{\gamma} = 10 \quad \text{to } 10^2 \text{ s}^{-1}$$

$$\text{Extrusion: } \dot{\gamma} = 10^2 \text{ to } 10^3 \text{ s}^{-1}$$

$$\text{Injection moulding: } \dot{\gamma} = 10^3 \text{ to } 10^4 \text{ s}^{-1}$$

The commercially available block copolymers are tailor-made to give optimal processing and mechanical properties for given applications. The S–B–S and S–I–S materials (compounds) commercially available can all be processed by conventional thermoplastic machinery[92] (see Chapter 4). The rheological properties of these materials are nearly equivalent to monomolecular melts of homopolymers at high shear rates. Many commercially available materials are, however, rather complex compounds.

Many of the newer commercially available thermoplastic elastomers comply very neatly with the rule stating that optimal properties might be expected for block copolymers where $|\delta_A - \delta_B|$ is close to zero and at least one component is crystalline. The thermoplastic polyesters (Hytrel) and some polyurethanes do give nearly monomolecular melt flow properties at temperatures above the melting point of the crystalline (hard) blocks (segments). These polymers constitute cases of the $A_n B_n$-type of block copolymers.

Polyolefin thermoplastic elastomers may be included in this class of materials even though the commercially available materials are rather complex compounds often consisting of mechanical mixtures of homopolymers with segmented multi-block copolymers. These products behave as monomolecular melts at least at the shear rates encountered during processing.

REFERENCES

1. van Oene, H. (1978). Rheology of polymer blends and dispersions. In: *Polymer Blends*, Vol. 1, (Ed. D. R. Paul and Seymor Newman), Academic Press, New York, Ch. 7.

2. Han, C. D. (1981). *Multiphase Flow in Polymer Processing*, Academic Press, New York.
3. Arnold, K. R. and Meier, D. J. (1970). *J. Appl. Polym. Sci.*, **14**, 427.
4. van Oene, H. (1972). *J. Colloid. Interfac. Sci.*, **40**, 448.
5. Han, C. D. and Funatsu, K. (1978). *J. Rheol.*, **22**, 113.
6. Chin, H. B. and Han, C. D. (1979). *J. Rheol.*, **23**, 557.
7. Chin, H. B. and Han, C. D. (1980). *J. Rheol.*, **24**, 1.
8. Vinogradov, G. V., Yacob, M., Tsebrenko, M. V. and Yudin, A. V. (1974). *Int. J. Polym. Metals*, **3**, 99.
9. Plochocki, A. P. (1978). In: *Polymer Blends*, Vol. 2, (Ed. D. R. Paul and S. Newman), Academic Press, New York, p. 319.
10. Walters, K. (1975). *Rheometry*, J. Wiley & Sons, New York.
11. Vinogradov, G. V. and Malkin, A. Ya. (1980). *Rheology of Polymers*, Springer Verlag, Berlin, p. 410.
12. Coleman, B. D., Markowitz, H. and Noll, W. (1966). *Viscometric Flow of Non-Newtonian Fluids*, Springer Verlag, Berlin.
13. Byron Bird, R., Armstrong, R. C. and Hassager, O. (1977). *Dynamics of Polymer Liquids*, Vol. 1, John Wiley & Sons, New York.
14. Lyngaae-Jørgensen, J. (1980). Relations between structure and rheological properties of concentrated polymer melts, Report issued by the Instituttet for Kemiindustri, The Technical University of Denmark, DK-2800 Lyngby, Denmark.
15. Alle, N. and Lyngaae-Jørgensen, J. (1980). In: *Rheology*, Vol. 2, (Ed. G. Astarita, G. Marucci and L. Nicolaus), Plenum Press, New York, p. 521.
16. Wilkes, G. L. and Samuels, S. L. (1973). Block and graft copolymers', Proc. 19th Sagamore Army Mater. Res. Conf., pp. 225–77.
17. Pico, E. R. and Williams, M. C. (1977). *Polym. Eng. & Sci.*, **17**, 573.
18. Wales, J. L. S. (1976). *The Application of Flow Birefringence to Rheological Studies of Polymer Melts*, Delft University Press.
19. Wang, F. W. (1978). *Macromolecules*, **11**, 1198.
20. Soli, A. L. (1978). Oscillatory flow birefringence measurements of block-copolymers in solution, Thesis, Univ. Wisconsin, Madison, USA.
21. Folkes, M. J. and Keller, A. (1971). *Polymer*, **12**, 222.
22. Ghijsels, A. and Raadsen, J. (1980). *Pure and Appl. Chem.*, **52**(5), 3159, IUPAC Report.
23. Ghijsels, A. and Raadsen, J. (1980). *J. Polym. Sci. Phys.*, **18**, 397.
24. Vinogradov, G. V., Dreval, V. E., Malkin, A. Ya., Yanovsky, Yu. G., Barancheeva, V. V., Borisenkova, E. K., Zabugina, M. P., Plotnikova, E. P. and Sabsai, O. Yu. (1978). *Rheol. Acta*, **17**, 258.
25. Kraus, G., Naylor, F. E. and Rollman, K. W. (1971). *J. Polym. Sci.*, A-2, **9**, 1839.
26. Chung, C. I. and Gale, J. C. (1976). *J. Polym. Sci., Phys.*, **14**, 1149.
27. Futamura, S. and Meinecke, E. (1977). *Polym. Eng. & Sci.*, **17**, 563.
28. Futamura, S. (1975). Effect of chemical structure of center block on physical and rheological properties of ABA block-copolymers, Thesis, University of Acron.
29. Matzner, M., Noshay, A. and McGrath, J. C. (1973). *ACS, Div. Polym. Chem., Preprint*, **14**(1), 68.

30. Noshay, A. and McGrath, J. E. (1977). *Block-Copolymers*, Academic Press, New York, p. 408.
31. Gouinlock, E. V. and Porter, R. S. (1977). *Polym. Eng. Sci.*, 17, 535.
32. Chung, C. J. and Lin, M. J. (1978). *J. Polym. Sci., Phys.*, 16, 545.
33. Chung, C. J., Griesbach, H. L. and Young, L. (1980). *J. Polym. Sci., Phys.*, 18, 1237.
34. Widmaier, J. M. and Meyer, G. C. (1980). *J. Polym. Sci., Phys.*, 18, 2217.
35. Cogswell, F. N. and Hanson, D. E. (1975). *Polymer*, 16, 936.
36. Erhardt, P. F., O'Malley, J. J. and Crystal, R. G. (1970). In: *Block Copolymers*, (Ed. S. L. Aggarwal), Plenum, New York, p. 195.
37. Le-Khac Bi and Fetters, L.-J. (1976). *Macromolecules*, 9, 732.
38. Masuda, R. and Aroi, P. (1981). Review with 46 references according to *Chem. Abstr.*, 95-133375.
39. Masuda, T., Kitamura, M. and Onogi, S. (1980). *Kenshu-Kyoto Daigaku Nippon Kagaku Seu i Kenkyusho*, 37, 15; *Chem. Abstr.*, 95-188425.
40. Leary, D. F. and Williams, M. C. (1970). *J. Polym. Sci., B.*, 8, 335; *J. Polym. Sci., Phys.*, (1973). 11, 345; (1974). 12, 265.
41. Krause, S. (1970). *Macromolecules*, 3, 84.
42. Meier, D. J. (1969). *J. Polym. Sci.*, C26, 81.
43. Helfand, E. (1975). *Macromolecules*, 8, 552; (1976). *Rubb. Chem. Technol.*, 49, 237.
    Helfand, E. and Wasserman, Z. (1977). *Polym. Eng. Sci.*, 17, 582.
44. Bolotnikova, L. S., Belenikevich, N. G., Evseev, A. K., Panov, Yu. N. and Frenkel, S. Ya. (1980). *Vysokomol. Soedin., Ser. A.*, 22, 1842.
45. Osaki, K., Kim, B.-S. and Kurata, M. (1979). *Polym. J.*, 11, 33.
46. Watanabe, H. and Kotaka, T. (1980). *Nippon Reoroji Gakkaishi*, 8, 25.
47. Stoelting, J. (1980). *J. Phys. Chem.*, 120, 177.
48. Watanabe, H., Kotaka, T., Hashimoto, T., Shibayama, M. and Kawai, H. (1982). *J. Rheol.*, 26, 153.
49. Pico, E. R. and Williams, M. C. (1977). *J. Polym. Sci., Phys.*, 15, 1585.
50. Pico, E. R. and Williams, M. C. (1977). *Polym. Eng. Sci.*, 17, 573.
51. Osaki, K., Kim, B.-S. and Kurata, M. (1978). *Polym. J.*, 10, 353.
52. Cox, W. P. and Merz, E. H. (1958). *J. Polym. Sci.*, 28, 619.
53. Lyngaae-Jørgensen, J., Alle, N. and Marten, F. L. (1979). *Adv. Chem. Series*, 176, 541.
54. Holden, G., Bishop, E. T. and Legge, N. R. (1969). *J. Polym. Sci., C*, 26, 37.
55. Zoteyev, N. P. and Bartenov, G. M. (1978). *Vysokomol. Soedin*, A20, 1781.
56. Enyiegbulam, M. and Hourston, D. J. (1981). *Polymer*, 22, 395.
57. Masuda, T., Matsumoto, Y. and Onogi, S. (1980). *J. Macromol. Sci., Phys.*, B17, 265.
58. Kotamura, M., Ishida, M., Masuda, T. and Onogi, S. (1981). *Nippon Reoroji Gakkaishi*, 9, 70; *Chem. Abstr.*, 95-133739.
59. Leblanc, J. L. (1976). *Rheol. Acta*, 15, 654.
60. Leblanc, J. L. (1976). *Polymer*, 17, 235.
61. Han, D. R. and Rao, D. A. (1979). *J. Appl. Polym. Sci.*, 24, 225.
62. Kraus, G. and Gruver, J. T. (1967). *J. Appl. Polym. Sci.*, 11, 2121.
63. Railsback, H. E. and Kraus, G. (1969). *Kautschuk und Gummi-Kunststoffe*, 22(9), 497.

64. Al'tzitser, V. S., Kandyrin, L. B., Anfimov, B. N. and Kuleznev, V. N. (1979). *Kuch. Rezina*, No. 12, 16.
65. Nguen Vin Chii, Isayev, A. I., Malkin, A. Ya., Vinogradov, G. V. and Kirchevskaya, I. Yu. (1975). *Vysokomol. Soedin*, **A17**, 855.
66. Sarakuz, O. N., Timchenko, B. N., Sinaiskii, A. G. and Makhmurov, A. G. (1979). *Plast. Massy*, No. 10, 37.
67. Ferrando, G. and Diani, E. (1976). *Technol. Plast. Rubber Interface*, 2nd Euro. Conf. Plast. Rubber Inst., (Prepr.), p. 1.
68. Golubkov, A. G., Zuppa, P. Ya. and Cherkasova, L. A. (1976). *Sint. Svoistva Uretanovykh Elastomerov*, **176–9**, 130, 137.
69. Vocel, J. and Stepankova, L. (1975). *Vodohospod. Cas.*, **23**, 268, *Chem. Abstr.*, 83-193889.
70. Akutin, M. S., Andrianova, B. V., Kulyamin, V. S., Zisman, D. O. and Babanova, L. A. (1975). *Plast. Massy*, No. 4, 44.
71. Paul, D. R., St. Lawrence, J. E. and Troell, J. H. (1970). *Polym. Eng. Sci.*, **10**, 70.
72. Kotaka, T. and White, J. L. (1973). *Trans. Soc. Rheol.*, **17**, 587.
73. Kim, B. S., Osaki, K. and Kurata, M. (1976). *Nippon Reoroji Gakkaishi*, **4**, 16.
74. Masuda, T., Matsumoto, Y., Matsumoto, T. and Onogi, S. (1977). *Nippon Reoroji Gakkaishi*, **5**, 135.
75. Osaki, K., Kim, B.-S. and Kurata, M. (1978). *Bull. Inst. Chem. Res., Kyoto Univ.*, **56**, 56.
76. Nemoto, N., Okawa, K. and Odani, H. (1973). *Bull. Inst. Chem. Res., Kyoto Univ.*, **51**, 118.
77. Kim, B. S. *Chem. Abstr.*, 95-133502.
78. Münstedt, H. (1975). *Angew. Makromol. Chem.*, **47**, 229.
79. Henderson, C. P. and Williams, M. C. (1979). *J. Polym. Sci., Lett.*, **17**, 257.
80. Bondi, A. (1968). *Physical Properties of Molecular Crystals. Liquids and Glasses*, John Wiley and Sons, New York, p. 172.
81. Brandrup, J. and Immergut, E. H. (1966). *Polymer Handbook*, Interscience, New York.
82. Vinogradov, G. V. (1977). *Polymer*, **18**, 1275.
83. Vinogradov, G. V. (1978). *J. Polym. Sci., Lett.*, **16**, 433.
84. Vinogradov, G. V., Malkin, A. Y. and Volosevitch, V. V. (1975). *J. Appl. Polym. Sci., Appl. Polym. Symposia*, **27**, 47.
85. Friedman, E. M. and Porter, R. S. (1975). *Trans. Soc. Rheol.*, **19**, 493.
86. Casale, A., Mononi, A. and Civardi, C. (1971). *Angew. Makromol. Chem.*, **53**, 277.
87. Casale, A. and Porter, R. S. (1971). *J. Macromol. Sci., Revs. Macromol. Chem.*, **C5**, 387.
88. Stazielle, C. and Benoit, H. (1971). *Pure & Appl. Chem.*, **26**, 451.
89. Kraus, G. and Gruver, J. T. (1965). *Trans. Soc. Rheol.*, **9**, 17.
90. Gruver, J. T. and Kraus, G. (1964). *J. Polym. Sci., A.*, **2**, 797.
91. Pezzin, G. (1962). *Materie Plastiche ed Elastomerie*, (August) English Translation: Instron Corporation, Canton, Mass.
92. Walker, B. M. (Ed.), (1979). *Handbook of Thermoplastic Elastomers*, Van Nostrand Reinhold, New York.

93. Denson, C. D. (1973). *Polym. Eng. Sci.*, **13**, 125.
94. Walters, K. (Ed.), (1982). *Rheometry: Industrial Applications*, Research Studies Press, Chichester.
95. Dealey, J. M. (1982). *Rheometers for Molten Plastics*, Van Nostrand Reinhold, New York.
96. Roe, R.-J. (1979). *Adv. Chem. Series*, **176**, 599.
97. Hansen, C. M. (1967). *J. Paint Technol.*, **39**, 104.
98. Hansen, C. M. (1967). 'The three-dimensional solubility parameter and solvent diffusion coefficients,' Thesis, The Technical University of Denmark.
99. Hansen, P. J., Hugenberger, G. S. and Williams, M. C. (1982). Proceedings 28th IUPAC Macromolecular Symposium, Amherst, p. 781.
100. Hansen, P. J. and Williams, M. C. (1982). SPE NATEC, Miami, p. 268, October 25–27.
101. Kotaka, T. and Watanabe, H. (1982). *Nihon Reoroji Gakkaishi*, **10**, 23.
102. Kitamura, M., Ishida, M., Masuda, T. and Onogi, S. (1981). *Nihon Reoroji Gakkaishi*, **9**, 70.
103. Masuda, T., Toei, K., Kitamura, M. and Onogi, S. (1981). *Nihon Reoroji Gakkaishi*, **9**, 77.
104. Taylor, G. I. (1934). *Proc. Roy. Soc.*, A, **146**, 501.
105. Cox, R. G. (1969). *J. Fluid. Mech.*, **37**, 601.
106. Goldsmith, H. L. and Mason, S. G. (1967). In: *Rheology*, Vol. 4, (Ed. F. R. Eirich) Academic Press, New York, Ch. II.
107. Tsebrenko, M. W., Rezanova, N. M. and Vinogradov, G. V. (1980). *Polym. Eng. Sci.*, **20**, 1023.
108. Ramachandran, S. and Christensen, E. B. (1983). *J. Non-Newtonian Fluid Mech.*, **13**, 21.
109. Ekong, E. A. and Jayaraman, K. (1984). *J. Rheol.*, **28**, 45.

APPENDIX

A.1. A Hypothesis for the Transition to a Monomolecular Melt State

A criterion for a transition from a two-phase state to a monomolecular melt state may be formulated as follows.[53] The total change in free energy (ΔG_T) by removing one domain from a melt at shear stress, τ, consists of two contributions:

$$\Delta G_T = \Delta G_{mix} + \Delta G_2 \qquad (A.1)$$

where ΔG_{mix} corresponds to the change in the free energy of mixing pure phases. Since the systems considered are originally two-phase systems (for $\tau = 0$), ΔG_{mix} is always positive. The action of a domain in a polymer melt is assumed to be equivalent to the action of a giant crosslink in a rubber. Removing one 'crosslink' is accompanied by a negative free energy change (ΔG_2).

Consider a system with U_A repetition units and a total volume fraction, v_A, of monomer A and U_B repetition units of monomer B. Since the free energy is a function of state the mixing operation may be performed by adding differential amounts of polymer A to a fixed amount of polymer B. The differential change in the total free energy change is then

$$d\Delta G_T = \left(\frac{\partial \Delta G_2}{\partial U_A}\right)_{T,P,U_B} \cdot dU_A \tag{A.2}$$

where

$$\left(\frac{\partial \Delta G_T}{\partial U_A}\right)_{T,P,U_B} = \left(\frac{\partial \Delta G_{mix}}{\partial U_A}\right)_{T,P,U_B} + \left(\frac{\partial \Delta G_2}{\partial U_A}\right)_{T,P,U_B} \tag{A.3}$$

If all differential contributions are negative the total free energy of mixing will be negative. To formulate a criterion for the destruction (disappearance) of the last domain a limiting expression for $(\partial \Delta G_T / \partial U_A)_{T,P,U_B}$ is elvaluated. The elastic contribution $(\partial \Delta G_2 / \partial U_A)_{T,P,U_B}$ is a decreasing function of the number of domains, while $(\partial \Delta G_{mix} / \partial U_A)_{T,P,U_A}$ is independent of the number of domains and for block copolymers[53] has a maximum and a minimum value.

The limiting form of $(\partial \Delta G_2 / \partial U_A)_{T,P,U_B}$ is approximately independent of v_A.

The critical condition for a transition to a monomolecular melt state with mixing ratio $v_{A,1}$ will be that

$$\left(\frac{\partial \Delta G_T}{\partial U_A}\right)_{T,P,U_B} \leq 0$$

in the interval $0 < v_A < v_{A,1}$. $\tag{A.4}$

$(\partial \Delta G_{mix} / \partial U_A)_{T,P,U_B}$ is found from Krause's expression for ΔG_{mix}. Krause[41] in a thermodynamic treatment of the Gibbs free energy change of mixing on domain destruction for the whole system of N_c copolymer molecules, occupying volume V, found

$$\Delta G_{mix}/kT = (V/V_r)v_A v_B \chi_{AB}(1 - 2/z) + N_c \ln(v_A^{v_A} v_B^{v_B}) - 2N_c(m-1)(\Delta S_{dis}/R) + N_c \ln(m-1) \tag{A.5}$$

where V is the total volume of the system, V_r is the volume of a lattice site, v_A and v_B are the volume fractions of monomer A and B in the copolymer molecule, respectively, z is the co-ordination number of the lattice, χ_{AB} is the interaction parameter between A units and B units, m is the number of blocks in the block copolymer molecule, $\Delta S_{dis}/R$ is the disorientation entropy gain on fusion per segment of polymer, k is the Boltzman constant, and T is the absolute temperature. On the differentiating eqn (A.5) with respect to polymer repeat units in a domain (U_A),

$$\left(\frac{\partial \Delta G_{mix}}{\partial U_A}\right)_{T,P,U_B} = \frac{V_A}{\bar{V}}RT\left\{\chi_{AB}\left(1-\frac{2}{z}\right)v_B^2\right.$$

$$+\frac{1}{\chi_n}\left[v_B\ln\frac{v_A}{v_B}+\ln\left(v_A^{v_A}v_B^{v_B}\right)\right]$$

$$\left.-\frac{2}{\bar{X}_n}(m-1)\frac{\Delta S_{dis}}{R}+\frac{1}{\bar{X}_n}\ln(m-1)\right\} \qquad (A.6)$$

For a copolymer with molar volumes, V_A, V_B and mole fractions n_A and n_B, $\bar{V}=n_AV_B+n_BV_B$, v_A is the volume fraction of repeat units found in domains, $v_B=1-v_A$ and $\bar{X}_n$ is the degree of polymerization.

$\bar{X}_n$ is defined as the number average of the ratio between the molar volume of polymer and the average molar volume $\bar{V}$. $(\partial \Delta G_2/\partial U_A)_{P,T,U_B}$ is given by the equation[14]

$$\left(\frac{\partial \Delta G_2}{\partial U_A}\right)_{P,T,U_B} = -\frac{\tau^2}{QT} \qquad (A.7)$$

for

$$\frac{3\eta_0 M_c H\rho}{c^2 RT}\dot{\gamma}>1$$

where τ is the shear stress (τ_{21}), T is the absolute temperature and

$$Q=k\frac{c^5 R}{\rho^4\bar{V}\bar{M}_c H^2} \qquad (A.8)$$

ρ is the melt density, c is the polymer concentration, R is the gas constant, $\bar{V}$ is the average molar volume of repeat units, $\bar{M}_c$ is twice the molecular weight between entanglements, H is the heterogeneity: $H=\bar{M}_w/\bar{M}_n$, and k is a dimensionless constant (which in Ref. 14 is estimated as 6×10^{-2}).

For

$$\left(\frac{\partial \Delta G_T}{\partial U_A}\right)_{T,P,U_B} = 0 \qquad (A.4)$$

substituting eqns (A.6), (A.7) into eqn (A.4) and simplifying, the following expression is found:

$$\tau_{cr}^2 = AT^2 + BT \tag{A.9}$$

$$A = Q\frac{V_A}{\bar{V}}\left\{ Rv_B^2\left(1 - \frac{2}{z}\right)\alpha + R/\bar{X}_n\left[\ln v_A^{*\,v_{A^*}}\, v_B^{*\,v_{B^*}} + v_B^*\ln\frac{v_A^*}{v_B^*}\right]\right.$$

$$\left. - (2R/\bar{X}_n)(m-1)(\Delta S_{dis}/R) + \frac{R}{\bar{X}_n}\ln(m-1)\right\} \tag{A.10}$$

$$B = Q\frac{V_A}{\bar{V}}Rv_B^{*2}\left(1 - \frac{2}{z}\right)\cdot\beta \tag{A.11}$$

for

$$\chi_{AB} = \alpha + \beta/T \tag{A.12}$$

where α and β are constants, τ_{cr} is the shear stress where the last domains disappear, and

$$v_B^* = 0\cdot5 + 0\cdot5\left/\left(1 - \frac{1}{2(\alpha + (\beta/T))\left(1 - \frac{2}{z}\right)\bar{X}_n}\right)\right. \tag{A.13}$$

for

$$v_B > 0\cdot5 - 0\cdot5\left/\left(1 - \frac{1}{2(\alpha + (\beta/T))\left(1 - \frac{2}{z}\right)\bar{X}_n}\right)\right. \tag{A.14}$$

where v_B^* corresponds to a maximum in

$$\left(\frac{\partial\Delta G_T}{\partial U_A}\right)T, P, U_B$$

In the calculations given below α in eqn (A.12) is taken equal to zero[96] and

$$\beta = \frac{\bar{V}}{R}(\delta_A - \delta_B)^2 \tag{A.15}$$

where δ_A and δ_B are solubility parameters for polymer A and polymer B, respectively. For polymers with strong polar and/or hydrogen bonding character, a better approximation for β may be obtainable by use of dispersion, polar and hydrogen bonding contributions following Hansen:[97,98]

$$(\delta_A - \delta_B)^2 = (\delta_A - \delta_B)_d^2 + (\delta_A - \delta_B)_p^2 + (\delta_A - \delta_B)_h^2$$

or

$$\beta = \frac{\bar{V}}{R} \sum_i^3 (\delta_A - \delta_B)_i^2 \tag{A.16}$$

In the limiting case $\bar{X}_n \to \infty$, $\alpha = 0$ and β is given by eqn (A.15). Equation (A.9) can be written

$$\tau_{cr} = \left(k \cdot \rho \frac{V_A}{\bar{V}} \cdot \frac{RT}{M_c} \left(1 - \frac{2}{z} \right) \right)^{1/2} \frac{|\delta_A - \delta_B|}{H} \tag{A.17}$$

Both eqns (A.9) and (A.17) are independent of volume fraction in accordance with the data of Arnold and Meier,[3] which did not reveal an influence on transition for $0.28 < v_A < 0.47$. The relationship to $|\delta_A - \delta_B|$ is in qualitative agreement with the observations of Futamura[27,28] and Matzner and co-workers.[29,30]

Equation (A.9) contains two parameters z and $\Delta S_{dis}/R$ which may be considered to be adjustable parameters. Krause[41] reports that z should vary between 4 and 12 and that Bondi[80] gives estimates for $\Delta S_{dis}/R$ in the range 0.9 to 4.3. Krause argues that $\Delta S_{dis}/R \approx 1$ and z close to 8 should be reasonable estimates.

A.1.1. Predictions in the S–B–S Case

The estimates used in the calculations are $k = 6 \times 10^{-2}$, $\delta_{PS} = 9.05$ (cal cm^{-3})$^{1/2}$, $\delta_{PB} = 8.30$ (cal cm^{-3})$^{1/2}$, the solubility parameters are estimated as an average of values taken from Brandrup and Immergut,[81] $\rho_{PS} = 1.04$ g cm^{-3}, $\rho_{PB} = 0.91$ g cm^{-3} $M_c^{PS} = 32\,000$, $M_c^{PB} = 5600$, $M_u^{PS} = 104$ g cm^{-3}, $M_u^{PB} = 54$ g mol^{-1}, $v_A = 0.25$, $v_B = 0.75$, $R = 8.3177 \times 10^7$ erg K^{-1} mol^{-1}, and $H \simeq 1.0$. The mixing rules

$$\frac{\rho_M}{M_{c,M}} = v_A \frac{\rho_A}{M_{c,A}} + v_B \frac{\rho_B}{M_{c,B}} \tag{A.18}$$

$$\bar{V} = n_A V_A + n_B V_B \tag{A.19}$$

gives $\bar{V}^{SBS} = 64.6 \, cm^3 \, mol^{-1}$, $M_{u,M} = 60.5$ and $\bar{x}_n = 57\,000/60.5 = 940$ for $\bar{M}_n = 5.7 \times 10^4$; M_u is the molecular weight per repetition unit. The limiting case described by eqn (A.17) gives an estimate of $2.9 \times 10^6 \, dyn \, cm^{-2}$ for the critical stress where a transition to monomolecular melt state is predicted. As discussed below this means that a monomolecular melt of SBS samples with high molecular weights should not be obtainable during normal processing operations because melt fracture will take place before or concurrent with such a stress level being reached. However, for low molecular weight samples and especially near the critical transition temperature, T_S, a monomolecular melt state may nevertheless be realized.

According to the argumentation presented above we can find T_S by specifying that the maximum value of

$$\left(\frac{\partial \Delta G_{mix}}{\partial U_A} \right)_{T,B,U_B} = 0 \qquad (A.20)$$

Thus, for a fixed value of z, $\Delta S_{dis}/R$ is fitted to eqn (A.20) in such a way that a transition temperature, T_S, equal to $150°C$ is obtained for an S–B–S sample with $\bar{M}_n = 57\,000$ and a styrene content of 25.4% by weight. This transition temperature was reported by Chung et al.[33] for an S–B–S sample with this molecular structure. For $z = 4$ one finds $\Delta S_{dis}/R = 3.9$. This value is at the high end of the interval given above (0.9–4.3). Thus the high value of $\Delta S_{dis}/R$ and the low value of v_A^* points to the conclusion that the transition to a monomolecular melt state is very sensitive to variations in degree of polymerization in the range where eqn (A.9) is nearly fulfilled.

A.1.2. Other Cases, a PS–PMMA Diblock Copolymer

In the limiting case $\bar{X}_n \to \infty$ and $\alpha = 0$ (eqn (A.17)), or when $BT \gg AT^2$, a nearly constant transition stress is predicted. The case where $BT \gtrsim AT^2$ and A is negative predicts decreasing transition stress for increasing temperatures in qualitative agreement with observations on a diblock copolymer of PS–PMMA.[53] In this last case calculation of the critical transition stress, τ_{cr}, is obtained as a difference between two large numbers. Consequently, such a calculation is encumbered with very large uncertainties and should only be expected to give order of magnitude estimates.

Experimentally, the critical stress was estimated as a point where an abrupt change of slope could be detected in the flow curves as shown on

Fig. A.1. Corresponding to the parts of the curves representing shear rates higher than τ_{cr}, Ferry's equation, $(1/\eta)=(1/\eta_0^*)+b\tau$, has been used to define a zero shear viscosity. Such plots are linear. Since the data for the styrene/methyl methacrylate copolymer show that τ_{cr} occurs very close to the Newtonian range (Fig. 11), this empirical procedure should

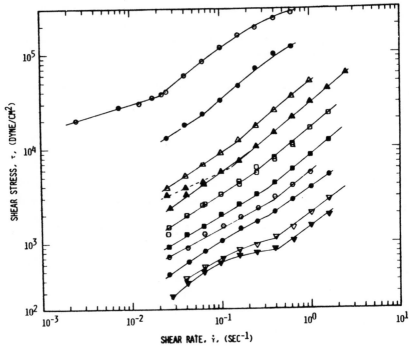

FIG. A.1. Double logarithmic delineation of shear stress vs. shear rate with temperature as discrete variable: ○, 160°C; ●, 170°C; △, 180°C; ▲, 190°C; □, 200°C; ■, 210°C; ◔, 220°C; ◆, 230°C; ▽, 240°C; ▼, 250°C. Material: S–M diblock copolymer.

give reliable values of zero shear viscosity. The zero shear viscosity estimated from data above this critical shear stress followed the Williams, Landel and Ferry (WLF) equation in the temperature range 160–200°C.[53]

Furthermore, strings of diblock copolymer extruded through an Instron capillary instrument at conditions considerably above the critical shear stress τ_{cr} and rapidly cooled under extensional flow conditions did show a significant change in morphology as shown from ultramicrotonic cuts near the periphery of the extruded strings.[14]

Since it is postulated that the melt state under capillary flow is a monomolecular melt state above τ_{cr}, one might be able to at least partly freeze this homogeneous state. However, the dramatic change in τ_{cr} with temperature which we observed for a PMMA–PS diblock copolymer with $\bar{M}_n = 500\,000$, $H = 1\cdot8$ and $25\,\text{wt}\%$ PMMA could not be simulated with constant A and B values in eqn (A.9). In fact τ_{cr} varied as $(\eta_0^*)^{1/2}$, where η_0^* is the viscosity estimated for a monomolecular melt state. This observation was rationalized in terms of 'mechanical domain instability'.

CHAPTER 4

Block Copolymers and Blends as Composite Materials

R. G. C. ARRIDGE

H. H. Wills Physics Laboratory,
University of Bristol, UK

and

M. J. FOLKES

Department of Materials Technology,
Brunel University, Uxbridge, UK

1. INTRODUCTION

The microstructure of regular block copolymers has already been discussed in Chapters 1 and 2. It is clear that these materials can be considered as micro-composites for they are two-phase and as such, it should be possible to predict their properties using traditional composite mechanics. Similarly, when block copolymers are blended with homopolymers, as is commonly carried out commercially, the resulting blend is also a composite, albeit a rather complex one. The direct application to block copolymers and blends of ideas developed for predicting the properties of conventional composite materials is a relatively new area of study, but is one that has considerable potential. Indeed, such an approach is particularly timely in view of the increasing use of blends for tailoring the properties of homopolymers. Accordingly, in this chapter, we briefly survey the mechanics of composites and then apply some of these theoretical ideas to the prediction of the mechanical properties of a block copolymer and its blend with polystyrene.

These materials have been processed in quite different ways, to give a range of microstructures. These include the comparatively simple 'single-crystal' block copolymer samples, discussed in Chapter 2, together with specimens that have been fabricated using normal technological processes such as injection moulding and screw extrusion.

The brief survey of composite mechanics is included so that the reader will not have to seek elsewhere for the origin of the formulae used in the sections concerned with the properties and microstructure of articles of practical use. The unity between the theories of composite materials and of block copolymers is illustrated by the applications but at this stage of the development of the subject only the more elementary aspects of composites theory are used. It is hoped that our account may prompt a more extensive activity.

2. THE MECHANICS OF COMPOSITES

2.1. Introduction

A composite material, as the term implies, is one made up from several components, each of different properties. These components may be amorphous or crystalline and may be of regular or irregular shape. Their distribution in the assembly may also be random or ordered, i.e. the assembly itself may possess a symmetry in space similar to that of a crystal or the various components may be randomly dispersed.

In the elastic analysis of composite materials we therefore have various tasks to perform: (1) To determine the 'unit cell', i.e. the smallest symmetry element in the assembly. (2) To discover the orientation of this unit cell at all points of the material. (3) To determine the concentration distribution function (the *spatial* description of the material). (4) To integrate the above information to derive the properties of the assembly. Such problems are found also in the fields of metallurgy, ceramics and petrology and the methods developed in these fields apply also to composites.

Now the 'unit cell' may be a crystal or it may itself be a composite so that, in general, we have to treat it as an elastic body which may have anisotropy. For example, the unit cell of semicrystalline polyethylene has orthorhombic symmetry with nine elastic constants, whereas the unit cell of a fibre composite with aligned fibres (Fig. 1) behaves like a material with transverse isotropy, requiring five elastic constants. (We shall see, in fact, that this model is appropriate for some block copolymers.)

If the elastic constants of the unit cell are known, therefore, or can be calculated, and if the orientation of the cell with respect to some 'global' scheme of axes is known, then the mean elastic constants of the whole assembly can be found, in principle, by integrating over all space. This can be done if we know the function which describes the orientation of

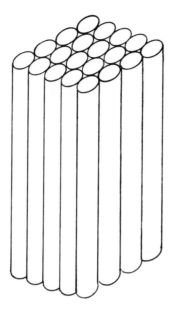

FIG. 1. The 'unit cell' of an aligned fibre reinforced composite exhibiting transverse isotropy.

the cell in the global scheme. This is usually termed the *orientation distribution function*. Figure 2 illustrates the orientation of a unit cell in global axes and describes this by the commonly used Euler angles (θ, ϕ, ψ). Then the orientation distribution function $\Phi(\theta, \phi, \psi) \sin\theta d\theta d\phi d\psi$ is defined as the probability that a unit cell has Euler angles in the solid angle $(\theta, \theta + d\theta; \phi, \phi + d\phi; \psi, \psi + d\psi)$. We must first transform the elastic constants of the unit cell from their own *local* axes of reference (usually the symmetry axes) to the global axes. To do this we require to do a tensor transformation (see, for example, Ref. 1) since the elastic constants are tensor quantities. Such transformations, though tedious, are not difficult and may be simplified in practice by using standard results. An example is given later of that for a transversely isotropic unit cell such as is found in fibre composites. We may refer to the transformed elastic constants as the global elastic constants of the unit cell. The mean properties of the assembly can now be found by integrating over all space, taking the orientation distribution function into account as well as the spatial distribution function. There are two common schemes. One is to sum over elastic stiffnesses, assuming uniform strain throughout the

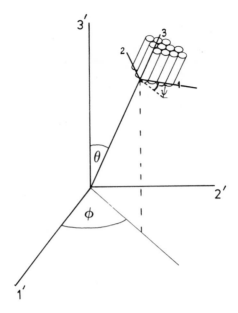

FIG. 2.   The orientation of a 'unit cell' as defined by local and global axes.

assembly; the other is to sum over elastic compliances assuming uniform stress. Neither method is wholly satisfactory but more precise methods have the disadvantage of being difficult to apply. (For a review see Ref. 2).

In some cases gross microstructure may be present and then the appropriate technique to employ is that of laminate theory, where simple weighted sums of elastic properties are used instead of the integral relations. Examples of this are given later.

If there is no symmetry present in the assembly, so that the phases are completely randomized, then only bounds on the elastic properties can be found, although in certain cases these may be fairly close.

2.2. Properties of the Unit Cell

For ordered arrays of fibres with unidirectional orientation, the properties of the unit cell can be calculated from knowledge of the properties of the constituents and their concentration.

It is convenient to use the two-suffix notation to describe the elastic constants. Then a general anisotropic material is described by a set of up to 21 constants (for triclinic symmetry) written as C_{ij} (stiffnesses) or S_{ij} (compliances). Their meaning is as follows. In a set of right-handed Cartesian axes (1, 2, 3) we define three tensile stresses σ_1, σ_2, σ_3 with corresponding strains e_1, e_2, e_3 and three shear stresses σ_4, σ_5, σ_6 with their corresponding strains e_4, e_5, e_6. Then

$$\sigma_k = \sum_{l=1}^{6} C_{kl} e_l, \ (k, \ l = 1, \ 2 \ \ldots \ 6) \text{ and } e_m = \sum_{p=1}^{6} S_{mp} \sigma_p$$

Matrix notation can be used so that we can write $\sigma = Ce$ and $e = S\sigma$ with $C = S^{-1}$ as matrix inverses of each other.

The transformation of axes referred to earlier requires tensor laws to be satisfied. It is not easy to formulate these transformations in a *general* way except by using tensors. This is done in the appendix. A specific case for hexagonal symmetry is given below and a tabulation of the changes in C_{ij} and S_{ij} for rotation about one of the principal axes is given in Hearmon.[1] Matrix schemes for transformation of axes when certain symmetries are present are given by Arridge[3] and one of these is also given below in Section 2.4.

2.3. The Elastic Constants for an Hexagonal Array of Cylinders

For such an array the symmetry requires five elastic stiffnesses. It is appropriate also for the case of transverse isotropy such as is found in fibre composites as well as in highly oriented polymers. Taking the fibre or cylinder axis as the 3-axis, the elastic constants required are C_{11}, C_{12}, C_{13}, C_{33}, C_{44} or their matrix inverses S_{11}, S_{12}, S_{13}, S_{33}, S_{44}. In the case where both fibres and matrix are isotropic the S_{ij} have been calculated.[4]

$$S_{11} = S_{22} = \frac{(1 - v_2)^2}{E_2} \left[\frac{2 + (\alpha_1 - 1) G_2 / G_1}{2 - v_1 + v_1 \alpha_2 + (1 - v_1)(\alpha_1 - 1) G_2 / G_1} - \frac{2 v_1 (1 - G_2 / G_1)}{v_1 + \alpha_2 + (1 - v_1) G_2 / G_1} \right]$$

$$S_{33}^{-1} = v_1 E_1 + (1 - v_1) E_2 + \frac{8 v_1 (1 - v_1) G_2 (v_1 - v_2)^2}{2 - v_1 + v_1 \alpha_2 + (1 - v_1)(\alpha_1 - 1) G_2 / G_1}$$

$$S_{13} = S_{23} = -S_{22} \left[v_2 - \frac{4 v_1 (1 - v_2)(v_2 - v_1)}{2 - v_1 + v_1 \alpha_2 + (1 - v_1)(\alpha_1 - 1) G_2 / G_1} \right]$$

$$S_{12} = S_{33} \left[v_2 - \frac{v_1(1+\alpha_2)(v_2 - v_1)}{2 - v_1 + v_1 \alpha_2 + (1 - v_1)(\alpha_1 - 1)G_2/G_1} \right]^2$$

$$- \left(\frac{1+v_2}{E_2} \right) \left\{ v_2 + (1 - v_2)v_1 \left[\frac{\alpha_2 - 1 - (\alpha_1 - 1)G_2/G_1}{2 - v_1 + v_1 \alpha_2 + (1 - v_1)(\alpha_1 - 1)G_2/G_1} \right. \right.$$

$$\left. \left. - \frac{2(1 - G_2/G_1)}{v_1 + \alpha_2 + (1 - v_1)G_2/G_1} \right] \right\}$$

$$S_{44} = \frac{1 - v_1 + (1 + v_1)G_2/G_1}{1 + v_1 + (1 - v_1)G_2/G_1} \cdot \frac{1}{G_2}$$

In these expressions E_i, G_i, v_i are, respectively, the Young's and shear moduli and the Poisson's ratio for component i, and $\alpha_i = 3 - 4v_i$. v_1 is the volume fraction of phase 1. The corresponding stiffnesses C_{ij} may be found by inverting the matrix $\{S_{ij}\}$. Modifications are required if the phases are anisotropic. Walpole[5] has treated theoretically the case of anisotropic phases and gives relations for bounds on the elastic constants and formulae for the self-consistent model. The reader is referred to the original papers for the expressions to use.

2.4. The Elastic Constants of a Composite

In order to find the overall elastic constants in a composite such as a block copolymer we need either the integral averaging procedure using an orientation distribution function (which gives the proportion of unit cells having orientations within a certain solid angle) or a summation scheme such as that used in laminate theory which is discussed in Section 2.5. For a transversely isotropic unit cell we have, because of the symmetry,

$$C_{66} = \tfrac{1}{2}(C_{11} - C_{12}), \quad C_{11} = C_{22}, \quad C_{13} = C_{23} \quad \text{and} \quad C_{44} = C_{55}$$

Then, for example, for $\langle C_{11} \rangle$, we find

$$\langle C_{11} \rangle = \iiint \Phi \sin \theta d\theta d\phi d\psi \left[(1 - \sin^2 \theta \cos^2 \phi)^2 C_{11} + \right.$$

$$\left. \sin^4 \theta \cos^4 \phi C_{33} + 2\sin^2 \theta \cos^2 \phi (1 - \sin^2 \theta \cos^2 \phi)(C_{13} + 2C_{44}) \right]$$

The integration over ψ involves ϕ only and introduces a numerical constant. If the global symmetry is that of isotropy about the global 3-

axis (fibre symmetry) then integration may be done over ϕ as well. This is not always possible but if it is we find, for $\langle C_{11} \rangle$

$$\langle C_{11} \rangle = 2\pi \int_0^\pi \sin\theta \Phi(\theta) d\theta \, \{ C_{11} + \sin^2\theta (C_{13} + 2C_{44} - C_{11})$$
$$+ \tfrac{3}{8}\sin^4\theta (C_{33} + C_{11} - 2C_{13} - 4C_{44}) \}$$

In the case of global fibre symmetry it is convenient to describe the distribution function $\Phi(\theta)$ in terms of an expansion in Legendre polynomials

$$\Phi(\theta) = \sum a_n P_n(\cos\theta)$$

where

$$a_n = \frac{2n+1}{2} \int_0^\pi \Phi(\theta) P_n(\cos\theta) \, d\theta = \frac{2n+1}{2} \langle P_n(\cos\theta) \rangle$$

For mechanical properties only two of the $P_n(\cos\theta)$ are needed.

$$P_2(\cos\theta) = \tfrac{1}{2}[3\cos^2\theta - 1]$$

and

$$P_4(\cos\theta) = \tfrac{1}{8}[35\cos^4\theta - 30\cos^2\theta + 3]$$

When the quantities P_n (which may be determined experimentally by techniques such as optical birefringence, NMR, infra-red dichroism, etc.) are used the means for the C_{ij} become expressible in terms of simple matrix quantities. For example, we have, for $\langle C_{11} \rangle$

$$\langle C_{11} \rangle = (a_0 \ a_2 \ a_4) \begin{bmatrix} \frac{8}{15} & \frac{1}{5} & 0 & \frac{4}{15} & \frac{8}{15} \\ \frac{8}{105} & -\frac{2}{35} & 0 & -\frac{2}{105} & -\frac{4}{105} \\ \frac{1}{105} & \frac{1}{105} & 0 & -\frac{2}{105} & -\frac{4}{105} \end{bmatrix} \begin{bmatrix} C_{11} \\ C_{33} \\ C_{12} \\ C_{13} \\ C_{44} \end{bmatrix}$$

for $\langle C_{12} \rangle$ the matrix is

$$\begin{bmatrix} \frac{1}{15} & \frac{1}{15} & \frac{1}{3} & \frac{8}{15} & -\frac{4}{15} \\ -\frac{2}{105} & -\frac{2}{105} & \frac{2}{15} & -\frac{2}{21} & \frac{8}{105} \\ \frac{1}{315} & \frac{1}{315} & 0 & -\frac{2}{315} & -\frac{4}{315} \end{bmatrix}$$

for $\langle C_{13} \rangle$

$$
\begin{bmatrix}
\frac{1}{15} & \frac{1}{15} & \frac{1}{3} & \frac{8}{15} & -\frac{4}{15} \\
\frac{1}{105} & \frac{1}{105} & -\frac{1}{15} & \frac{1}{21} & -\frac{4}{105} \\
-\frac{4}{315} & -\frac{4}{315} & 0 & \frac{8}{315} & \frac{16}{315}
\end{bmatrix}
$$

for $\langle C_{33} \rangle$

$$
\begin{bmatrix}
\frac{8}{15} & \frac{1}{5} & 0 & \frac{4}{15} & \frac{8}{15} \\
-\frac{16}{105} & \frac{4}{35} & 0 & \frac{4}{105} & \frac{8}{105} \\
\frac{8}{315} & \frac{8}{315} & 0 & -\frac{16}{315} & -\frac{32}{315}
\end{bmatrix}
$$

and, for $\langle C_{44} \rangle$

$$
\begin{bmatrix}
\frac{7}{30} & \frac{1}{15} & -\frac{1}{6} & -\frac{2}{15} & \frac{2}{5} \\
-\frac{1}{42} & \frac{1}{105} & \frac{1}{30} & -\frac{2}{105} & \frac{1}{35} \\
-\frac{4}{315} & -\frac{4}{315} & 0 & \frac{8}{315} & \frac{16}{315}
\end{bmatrix}
$$

If the distribution of unit cells is random in orientation then $a_0 = 1$, $a_2 = a_4 = 0$ and the $\langle C_{ij} \rangle$ are given by the first lines only of the above matrices (after the two right-hand matrices have been multiplied out).
Thus

$$
\langle C_{11} \rangle = \tfrac{8}{15} C_{11} + \tfrac{1}{5} C_{33} + \tfrac{4}{15} C_{13} + \tfrac{8}{15} C_{44}
$$

It will be found that only two independent elastic constants remain (as appropriate for an *isotropic* assembly). These are $\langle C_{11} \rangle$ and $\langle C_{12} \rangle$. For preferred orientations more elastic constants are required, and these may be found using the appropriate values of a_2 and a_4.

C_{11} and C_{12} are related to the familiar E, G and v of isotropic materials by the formulae

$$
E = C_{11} - \frac{2C_{12}^2}{C_{11} + C_{12}}
$$

$$
G = \tfrac{1}{2}(C_{11} - C_{12})
$$

$$
v = \frac{C_{12}}{C_{11} + C_{12}}
$$

2.5. The Laminate Theory
If the concentration distribution function of the constituents of the composite (whether or not these constituents are themselves composites)

is markedly 'granular' so that clear separation into grossly dissimilar phases is present, then it may be preferable to treat the assembly as a laminate. In laminates, such as multi-ply wood, or glass-fibre laminates, there are present a limited number of phases which may differ only in orientation or also in composition.

In Fig. 3 each laminate is shown as having different elastic properties. We need first of all to transform these *sheet* properties to properties referred to the global axes. This is done, as before, by tensor transformation. With respect to the *global* axes, therefore, let the kth layer have elastic stiffnesses ${}^kC_{ij}$. Then we may postulate two different cases — plane strain and plane stress — for which to calculate the overall properties.

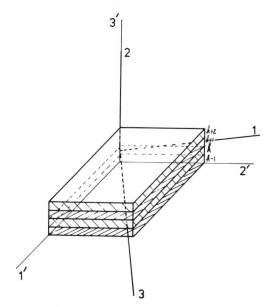

FIG. 3.   The definition of axes used in the laminate theory.

In plane strain we assume the displacements

$$u_1 = u_1(x_1, x_2)$$
$$u_2 = u_2(x_1, x_2)$$
$$u_3 = 0$$

Then $e_3 = e_4 = e_5 = 0$, so that $\sigma_4 = \sigma_5 = 0$ and $\sigma_3 = C_{31}e_1 + C_{32}e_2 + C_{36}e_6$.

If we wish to use the compliance matrix we need to write

$$R_{ij} = S_{ij} - \frac{S_{i3} S_{j3}}{S_{33}} \quad (i,j = 1,2,6)$$

and

$$e_i = \sum_j R_{ij} \sigma_j$$

In plane stress $\sigma_3 = \sigma_4 = \sigma_5 = 0$, $e_4 = e_5 = 0$ and $e_3 = S_{31}\sigma_1 + S_{32}\sigma_2 + S_{36}\sigma_6$. e_3 is not independent of e_1, e_2 and e_6 and may be eliminated. We write

$$\sigma_i = C_{ij} - \frac{C_{i3} C_{j3}}{C_{33}} \quad (i,j = 1,2,6)$$

$$= \sum_j Q_{ij} e_j$$

where Q_{ij} is the matrix of reduced stiffnesses.

For a lamina of thickness h the bending moments are defined (Timoshenko and Woinowsky-Krieger[6]) as

$$M_i = \int_{-h/2}^{h/2} \sigma_i z \, dz \quad (i = 1, 2, 6)$$

The stress resultant is defined as

$$N_i = \int_{-h/2}^{h/2} \sigma_i \, dz$$

Then, writing $e_i = e_i^0 + z\kappa_i$, where e_i^0 is the in-plane strain and $z\kappa_i$ the strain due to curvature κ_i we have, following Tsai[7]

$$\begin{bmatrix} N \\ M \end{bmatrix} = \begin{bmatrix} A & B \\ B & D \end{bmatrix} \begin{bmatrix} e \\ \kappa \end{bmatrix}$$

where $A_{ij} = \int_{-h/2}^{h/2} C_{ij} \, dz$ (plane strain), (Q_{ij} for plane stress)

$$B_{ij} = \int_{-h/2}^{h/2} C_{ij} z \, dz \quad \text{(plane strain), } (Q_{ij} \text{ for plane stress)}$$

$$D_{ij} = \int_{-h/2}^{h/2} C_{ij} z^2 \, dz \text{ (plane strain), } (Q_{ij} \text{ for plane stress)}$$

Now if the laminate is made up from n layers, each homogeneous and of global elastic constants ${}^kC_{ij}(k=1,\ldots n)$ (or, for plane stress, ${}^kQ_{ij}$) then the above integrals may be replaced by sums such as

$$A_{ij}=\sum_{k=1}^{n} {}^kC_{ij}(h_{k+1}-h_k)$$

$$B_{ij}=\tfrac{1}{2}\sum_{k=1}^{n} {}^kC_{ij}(h_{k+1}^2-h_k^2)$$

$$D_{ij}=\tfrac{1}{3}\sum_{k=1}^{n} {}^kC_{ij}(h_{k+1}^3-h_k^3)$$

where $h_{k+1}-h_k$ is the thickness of the kth layer.

An application of this theory to block copolymers follows in Section 4.2.

3. THE MECHANICS OF 'SINGLE-CRYSTAL' S–B–S

It was shown in 1970[8] that extruding a block copolymer such as S–B–S resulted in a highly ordered quasi 'single-crystal' structure. The samples obtained by the extrusion procedure were shown by X-ray and electron microscopy to consist of an hexagonal array of polystyrene cylinders embedded in a polybutadiene matrix, with the cylinder axes along the extrusion direction (see Chapter 2). Such a single-crystal sample of S–B–S copolymer is, in fact, a miniature fibre reinforced system and in two papers[9,10] the present authors compared mechanical measurements of the elastic properties of the material with theoretical predictions using fibre reinforcement theory.

Measurements of Young's modulus E_θ, using a dead-loading technique were made on samples cut with various orientations θ of the polystyrene cylinders to the direction of loading (see Fig. 4).

Now the compliance S'_{33} at any angle to the axis of symmetry of a material with transverse symmetry about the 3-axis is given by

$$S'_{33}=1/E_\theta=S_{11}\sin^4\theta+(2S_{13}+S_{44})\sin^2\theta\cos^2\theta+S_{33}\cos^4\theta \quad (1)$$

and Young's modulus E_θ measured at only three angles is sufficient to provide values for S_{11}, $(2S_{13}+S_{44})$ and S_{33}. The results from more than three were then used to provide refined values for these compliances using a least-squares technique. By this method the best-fit values for the

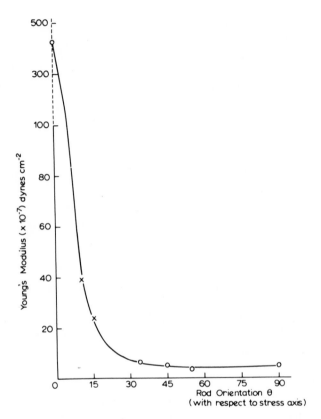

FIG. 4. The variation of Young's modulus with orientation angle θ (1 dyn cm^{-2} = 0·1 Nm^{-2}) for the S–B–S single-crystal sample of Kraton 102.

compliances S_{33}, S_{11} and $2S_{13} + S_{44}$ together with their associated errors were derived.

Strictly, it is necessary to determine S_{13} from measurement of the Poisson's ratio v_{13}. However, from physical considerations and experimental observation it was found that $v_{13} \ll 1$.

This arises because if a stress is applied along the 1-direction (perpendicular to the cylinders) the transverse contraction along the 3-direction will be at least two orders of magnitude smaller than that along the 2-direction, owing to the constraining effect of the fibres. This was found to be fully justified for our copolymer samples since the Young's modulus of the polystyrene rods was approximately 2GPa, whereas that of the polybutadiene matrix was about 2·5 MPa. This implies that $2S_{13} \ll S_{44}$ and consequently may be neglected in comparison with it.

From the least-squares analysis therefore, the constants derived were (units are in $N^{-1} m^2$)

$$S_{11} = (2·3 \pm 0·3) \times 10^{-7}$$
$$S_{33} = 2·4 \times 10^{-9}$$
$$S_{44} = (6·6 \pm 0·7) \times 10^{-7}$$

Measurements were also carried out at room temperature to determine the values of S_{44} and S_{66} directly by using a free oscillation torsion pendulum at a frequency in the range 0·1–1·0 Hz.

The method is a standard one and in the case of the S–B–S single-crystal material may give values of the shear compliances. For torsion about the 3-axis, the rigidity modulus $n = k_1 a^3 b / c S_{44}$, where a, b are, respectively, the dimensions of the sample in the 2- and 1-directions, c its length along the 3-direction and k_1, a constant dependent on the sample geometry, is given by Hearmon.[1]

$$k_1 = \frac{1}{3} \left\{ 1 - \frac{0·6274}{u} \sum_{j=0}^{\infty} \frac{1}{(2j+1)^5} \tanh \left(\frac{2j+1}{2} \pi u \right) \right\}$$

where $u = b/a$.

For torsion about the 1-axis, n is given by $n = k_1 c^3 a / b S_{66}$, k_1 has the same form as above but now $u = \frac{a}{c} \{ S_{66}/S_{44} \}^{1/2}$ and must be greater than 1 for the expression for n to be valid. It is important in this case that a/c should be as large as possible to ensure that, irrespective of the exact values of S_{44} and S_{66} occurring in the expression for u, k_1 is reasonably independent of aspect ratio. For the measurement of S_{66} the samples must therefore be thin sections cut *across* the extruded plug perpendicular to the extrusion direction. The following values were found initially for S_{44} and S_{66}:

$$S_{44} = (3·1 \pm 0·2) \times 10^{-7} (N^{-1} m^2)$$
$$S_{66} = (5·9 \pm 0·3) \times 10^{-7} (N^{-1} m^2)$$

The results obtained for S_{11}, S_{33} and S_{44} from the Young's modulus tests were in good agreement with predictions from fibre reinforcement theory, provided that a value of 0.37 for the Poisson's ratio of the butadiene phase was used rather than the more usual figure of 0·5 for a rubber.

The *overall* Poisson's ratio v_{12} for the S–B–S 'composite' was then predicted to be 0·64. If the *measured* values of S_{66} and S_{11} are, however, used to derive v_{12} (since $S_{66} = 2(S_{11} - S_{12})$) a much lower value of v_{12} is found. The measurements of S_{66}, however, like those of S_{44}, may be affected by the end effects discussed below. It was not possible on the very small samples cut for S_{66} measurements, to verify this and consequently the experimental S_{66} values are not used.

The values obtained for S_{44} from the fitting of Young's modulus values for samples cut at various angles θ to the symmetry axis and for S_{44} from torsion tests differed by a factor of 2. In the two papers cited this difference was discussed in great detail and found to be caused by the failure, for highly anisotropic materials, of the commonly assumed Principle of St. Venant.

This principle, as understood by most physicists or engineers, amounts to the statement that end effects in the torsion or tension of cylinders, or in the bending of beams, die away within a length approximately equal to the maximum cross-sectional dimension. A common application of the principle is to ensure that in making measurements of elastic constants in tension or in torsion the specimen length-to-width ratio is large enough for these end effects to be ignored. How large this ratio has to be becomes a matter for experiment, since the wise experimentalist will use samples of two or more different lengths to see whether the effect is discernible within experimental error. A common rule of thumb is to take a value of 10 for the length-to-width ratio—but this is nothing more than a convenient figure.

In the case of tensile tests, certain specimen shapes are recommended by standards institutions. These shapes have been developed by analysis of the stresses in bars under terminal loads and such studies[11] illustrate the practical truth of St. Venant's principle for simple geometries and for isotropic specimens. St. Venant in fact stated the principle as being valid for perfect cylinders only, not necessarily for those of arbitrary cross-section, and as Toupin[12] showed the stresses in a beam of dumb-bell cross-section loaded at one end by couples whose resultant is zero will persist far along the beam.

A criterion based upon elastic stored energy was proposed by Toupin and applied to anisotropic materials by Horgan.[13] This criterion leads to the concept of a *characteristic decay length* for the stored elastic energy and shows that the decay of the stress irregularity is an exponential function of the distance from the boundary.

For anisotropic materials Horgan showed that the decay length was of the form $\lambda = Cd(E/G)^{1/2}$, where d is the maximum lateral dimension of the sample and C is a constant. Now for isotropic materials E/G lies between 2·6 and 3·0, so that $\lambda \simeq 1·6 - 1·7\,Cd$. For very anisotropic materials, however, such as SBS copolymer in the 'single-crystal' form obtained by extrusion, the relevant modulus ratio is S_{44}/S_{33} and this lies between $3·1 \times 10^{-7}/2·4 \times 10^{-9}$, and twice this, depending upon the value of S_{44} used, i.e. between 130 and 260, so that λ may be as high as $16Cd$.

In such cases neglect of the stress distribution at the points of loading becomes quite unjustifiable, the 'rule-of-thumb' figure of 10 for the length-to-width ratio of the specimen totally inadequate, and only two alternatives remain. One is to determine the stress distribution for the particular method of sample clamping, solve the elastic equations of equilibrium and hence derive the elastic constants. The other is to choose a sample length/width ratio so large that the end effect may still be ignored, and this will have to be done, *ceteris paribus*, by trial and error.

The first alternative is extremely difficult and only likely to succeed in a few cases. The second alternative seems the better one provided sufficient material is available and this was the choice made by the authors in their studies of oriented S–B–S (and by Arridge et al. in the study of ultra-oriented polyethylene[14]).

4. TECHNOLOGICAL PROCESSING OF BLOCK COPOLYMERS AND BLENDS

So far in this chapter we have been concerned with the application of composite mechanics to simply oriented specimens of block copolymer. A distinction must be drawn, however, between studies where the method of sample fabrication is chosen to promote the development of a simple microstructure and those where the samples have been prepared using one of the normal polymer fabrication processes, e.g. injection moulding and extrusion. Considering the widespread conversion of block copolymers into moulded components it is perhaps surprising that so little has been published in the scientific literature concerning the relationships between processing, microstructure and properties. There is no reason to expect that the microphase morphology in block copolymers after processing will be identical to the equilibrium morphology as studied in simple laboratory specimens. Also, aside from the possible effects of complex thermal and flow fields on the morphology of block copolymers, further complications can arise owing to the fact that many applications for these materials demand significant modifications to their properties, which cannot be achieved by changes to the processing conditions alone. For this reason, the virgin copolymer is frequently blended with other additives, e.g. homopolymers, plasticizers, etc., to produce a compound having optimum processing and property characteristics. In view of the potentially complex microstructural changes that are expected to result from both the compounding and moulding processes, it is convenient to divide the discussion into two parts; one concerning the processing of the virgin block copolymer and the other concentrating on the effects of added homopolymer. In both cases, the use of simple composite mechanics to interpret the properties of the resulting products will be demonstrated.

4.1. Processing of S–B–S Block Copolymers

One of the earliest systematic studies using commercial fabrication technology was reported by Aggarwal.[15] In this case, the Shell copolymer Kraton 1101 was processed on a two-roll mill to yield a hide, which had a pronounced mechanical anisotropy, being much stiffer along the direction of rolling compared with the orthogonal direction. Transmission electron microscopy showed that the microphase morphology was cylindrical and that in the outside layers of the sheet these microphases were preferentially aligned along the rolling direction, as shown in Fig. 5. Of particular significance is the fact that the equilibrium cylindrical morphology is preserved in spite of the more demanding process to which this material was subjected compared with the method used for the preparation of single-crystal specimens.[16] The mechanical properties of the milled material were interpreted using composite mechanics and similar good agreement between theory and experiment was found, as in the case of the single-crystal specimens, discussed earlier in this chapter. Charrier and Ranchoux[17] studied the mechanical properties of two S–B–S block copolymers, Shell Kraton 1101 and Kraton 3226, the latter containing pigment and plasticizer. These materials were compression-moulded and later subjected to a shearing process in a rectangular channel. Two observations are particularly noteworthy; in the case of Kraton 1101 the material undergoes strain softening after the first extension during mechanical testing and the sheared sample shows anisotropic properties. The interpretation of the mechanical data assumed that the morphology of this particular material was not very different from the generally accepted morphology of solvent-cast films, i.e. polystyrene spheres in a polybutadiene matrix.[18,19] However, it is known that Kraton 1101 contains approximately 25% vol. fraction of

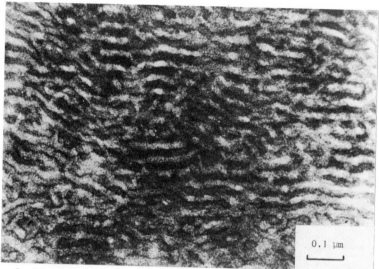

0.1 μm

FIG. 5. Transmission electron micrograph of a thin section cut longitudinally from a milled and compression-moulded sheet of Kraton 101. Reproduced from Ref. 15 by courtesy of the publisher, Butterworth & Co. (Publishers) Ltd ©.

polystyrene and a cylindrical microphase morphology would be expected in the equilibrium state. The proposed mechanism of strain softening, etc., as envisaged by Charrier and Ranchoux is incompatible with this observation unless the microphase morphology is grossly disrupted during the process of compression moulding. Mechanical anisotropy was also observed in compression-moulded samples of block copolymers, having polystyrene cylinders in a polybutadiene matrix, by Lewis and Price[20] and Harpell and Wilkes.[21]

Mieras and Wilson[22] have studied the anisotropy developed in block copolymers exhibiting a rod-like morphology. The anisotropy resulted from either screw extrusion or injection moulding. Melt processing of these materials produces a high degree of orientation and continuity of the cylindrical microphases, which leads to strongly anisotropic mechanical behaviour. Furthermore, this leads to pronounced yielding effects in those directions where the polystyrene phase is continuous. The effects are enhanced by annealing above ambient temperatures, or even storage at room temperature.

4.2. Microstructure and Mechanical Properties of Injection Mouldings

Injection moulding is one of the most important fabrication methods used throughout the plastics industry. The conditions to which a material is subjected are very severe in this process compared with the simple methods used for preparing laboratory samples. Figure 6 shows a schematic diagram of the injection-moulding cycle for a conventional screw machine. There are at least two parts of this cycle where one might anticipate that the equilibrium morphology of the block copolymer will

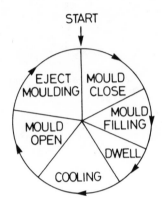

FIG. 6.   Schematic diagram showing the main stages involved in the injection-moulding cycle.

be severely disrupted, namely, during screw-back and injection. The rôle of the first of these processes is the production of a homogeneous melt and involves the transport and melting of the raw material by means of the screw rotation. When a sufficient volume of melt has been produced, the screw rotation stops and the screw then advances forward to inject the melt into the mould cavity. The time required to fill the mould completely can be less than 1 s and, consequently, the stresses acting on the melt will be very large and result in high shear rates, especially in the vicinity of the gate.

In view of these considerations, it is rather surprising that such pronounced anisotropy exists in melt processed samples when one might anticipate major disruption to the microphase morphology. In order to clarify this matter, Folkes and Nazockdast[23] have carried out a detailed investigation of the relationship between the microstructure and mechanical properties of injection-moulded plaques of Kraton 102. This is the same raw material as used in the earlier single-crystal work and provides an opportunity for applying the anisotropy data for the single-crystal specimens in the interpretation of the physical property data of the injection-moulded samples.

4.2.1. Moulding and Microstructure of the Plaques

The mould used by Folkes and Nazockdast was a two-impression edge-gated square plaque of dimensions $80 \times 80 \times 3$ mm. The moulding was conducted using a Daniels 350/120 injection-moulding machine with a 200 g shot weight capacity. A range of plaques were produced using a variety of moulding conditions. However, it was found that the microstructure of the plaques was very insensitive to comparatively large changes in moulding conditions. Throughout the moulding operation, the barrel temperature was maintained in the range 130–150°C with the mould nominally at ambient temperature, ~ 28°C. Fast injection speeds were employed.

As soon as the mouldings were ejected, a very pronounced mechanical anisotropy was apparent, as assessed by a simple flexural test. The microstructure of the moulded plaques was examined by cutting through-thickness sections from a number of locations, as shown schematically in Fig. 7. Particular attention was devoted to sections cut

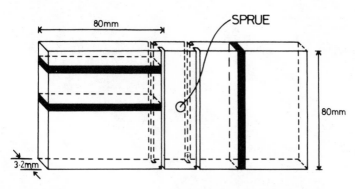

FIG. 7.   Schematic diagram of the two-impression plaque mould used for the injection-moulding studies.[23]

from along the axis of flow. When these sections were examined in the optical microscope using crossed-polars, it was found that a systematic extinction occurred in different layers through the section. This showed that the material was optically anisotropic and that the extinction angle varied through the section, i.e. the orientation was inhomogeneous. A

FIG. 8. Birefringence pattern observed in a thin section of S–B–S block copolymer Kraton 102, cut parallel to the major flow direction in the plaque mould.[23]

typical birefringence pattern obtained for one position of the section with respect to the polarizer is shown in Fig. 8. It can be seen that the pattern is very slightly asymmetric and this is due to a small difference in temperature on the two mould faces. The section appeared to consist of seven quite well defined birefringent layers. The extinction angle and birefringence were measured for each layer and the results are shown in Table 1. The birefringence is very close to the form birefringence value of

TABLE 1

Layer	Thickness (mm)	Extinction angle ($\theta°$)	Birefringence $\Delta n \times 10^4$	Young's modulus $\times 10^{-8} Nm^{-2}$
1	0·57	7	5·7	0·83
2	0·62	4	4·6	1·79
3	0·30	15	5·5	0·23
4	0·20	90 (axis)	2·2	0·04
5	0·57	15	5·5	0·23
6	0·62	4	4·6	1·79
7	0·30	7	5·7	0·83

5.15×10^{-4} for the single-crystal S–B–S samples studied by Folkes and Keller.[24] In view of this and the comparative invariance of the measured value through the section, it is reasonable to infer that the birefringence in the injection-moulded plaque is associated with localized alignment of polystyrene cylinders in a polybutadiene matrix. The arrangement of the polystyrene rods in the section is shown schematically in Fig. 9. The rods

Injection Direction

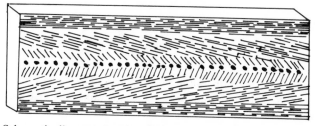

FIG. 9.   Schematic diagram showing the orientation of the polystyrene cylinders in a thin section cut parallel to the major flow direction in the plaque mould.[23]

are highly aligned parallel to the flow direction in the top and bottom layers of the section but become progressively inclined towards the axis of the section where they eventually exhibit a near random orientation distribution. The proportion of the highly aligned surface layers depends on moulding conditions, e.g. it increases when the melt temperature is reduced.

4.2.2. Mechanical Properties

As a first step towards an understanding of the relationship between the microstructure of the plaque and its mechanical properties, the tensile moduli of the sections cut from the complete moulding were measured. The work of Horgan[13] and Arridge and Folkes[10] has shown that a large sample aspect ratio (length/width) is required when measuring the tensile modulus of highly anisotropic materials. The aspect ratio of the samples cut from the plaque is limited by the dimensions of the moulding and it is not obvious, a priori, that a state of uniform strain will exist throughout the section during loading. Accordingly, it was necessary to confirm that this was the case by examining the deformation of a small grid applied to the sample surface. For the case of the samples cut from the axis of the mouldings, an average experimental value of tensile modulus equal to $(1 \cdot 14 \pm 0 \cdot 1) \times 10^8 \, \mathrm{Nm}^{-2}$ was obtained. This should be compared with the value of $4 \times 10^8 \, \mathrm{Nm}^{-2}$ obtained for the longitudinal tensile modulus of the single-crystal samples.

For a sample cut as above, the prediction of Young's modulus from a knowledge of the arrangement of the polystyrene cylinders in the section is facilitated by the existence of a state of uniform strain. It is appropriate, therefore, to regard the partially oriented material as a symmetrical laminate and use the laminate theory as discussed earlier in Section 2.5. There it was shown that if there is no curvature, the stress resultant N is given simply by

$$\mathbf{N} = \mathbf{A} \, \mathbf{e}$$

where the matrix

$$A_{ij} = \sum_{k=1}^{n} {}^{k}C_{ij}(h_{k+1} - h_{k})$$

As in Fig. 3 we take the common perpendicular of the laminate to be the 3'-axis and calculate the stress, and therefore the overall stiffness, along the 1' or 2' axes.

The assumption of *plane strain* in the laminate is not realistic because it requires stresses in the 3'-direction to maintain it. A more realistic assumption is that of *plane stress*. We can then write for the kth layer

$e_{1'} = {}^kS_{1'1'}\sigma_{1'}$, $e_{2'} = {}^kS_{2'1'}\sigma_{1'}$, $e_{3'} = {}^kS_{3'1'}\sigma_{1'}$, where the only stresses acting are asssumed to be in the $1'$ direction. In fact a laminate in elastic equilibrium will not have uniform stress or strain in it, since each layer is constrained by its neighbours. The state of stress is not simple and interlaminar shear may develop — see, for example, Puppo and Evensen.[25] For our purposes, however, we will assume the separate layers to be unconstrained by each other and assume also that the strain in the $1'$ direction is the same for each layer and is equal to e, say. Then the corresponding stress in each layer in the $1'$ direction is $\sigma_{1'} = e/{}^kS_{1'1'} = E'_k e$.

This leads to the simple 'rule of mixtures' law for the overall modulus, since the total force must be

$$\sigma_c t_c = \sum_{k=1}^{n} E'_k t_k$$

where

$$t_c = \sum_{k=1}^{n} t_k$$

and $t_k = h_{k+1} - h_k$ is the thickness of the kth layer. Hence, $E_c = \Sigma E'_k t_k / t_c$. $E'_k = 1/{}^kS_{1'1'}$ is the Young's modulus of the kth layer and this may be found from the global value of ${}^kS_{1'1'}$ for that layer, this in turn, being found by transformation of the 'local' ${}^kS_{ij}$.

The transformation is in fact simple in our case because the orientation of the local axis is such that the local 3-axis lies in the plane of the section cut from the plaque. Thus, if the orientation of this 3-axis is known with reference to the direction of loading, the value of ${}^kS_{1'1'}$ is easily found from the mechanical anisotropy data for the single-crystal samples.[9] In Table 1 the measured orientation angles are given for each layer in the section and the corresponding values of $E' = 1/S_{1'1'}$.

Hence, using these values of layer modulus in the equation for the overall modulus, derived above,

$$E_c = \sum_{k=1}^{n} E'_k t_k / t_c$$

a predicted value of $1{\cdot}04 \times 10^8 \, \mathrm{Nm}^{-2}$ for the section stiffness was obtained, in close agreement with the experimental value of $(1{\cdot}14 \pm 0{\cdot}1) \times 10^8 \, \mathrm{Nm}^{-2}$. Similar agreement has also been obtained for other sections cut from the plaque, showing that this approach provides an unusually simple method of analysing the mechanical properties of mouldings resulting from a commercially important fabrication process.

4.3. Development of Composite Microstructure during Moulding

There remain two important points related to the development of the microstructure in the block copolymer mouldings. The first is the somewhat surprising result that the mechanical anisotropy is explicable in terms of a model in which the polystyrene rods are continuous, i.e. no major change has occurred in the microphase morphology during injection moulding. This result is incompatible with previous work concerning the micro-rheology of block copolymers[26] and demands closer examination. The second point will consider how the final orientation in the moulded component is related to the flow processes occurring during the filling of the mould cavity.

The rheological properties of S–B–S copolymers are unusual compared with homopolymers. Based on extensive viscosity data, Arnold and Meier[26] have suggested that these copolymers can exist in three distinct rheological states depending on shear rate. At low shear rates, the three-dimensional molecular network structure remains essentially intact. At intermediate deformation rates, the three-dimensional domain network will be disrupted leading to domain aggregates, but these are not linked together to form a three-dimensional network — see Fig. 10. Finally, at high deformation rates, these aggregates will in turn be disrupted and the melt will behave as an assemblage of individual, non-aggregated molecules, i.e. the two-phase microstructure will be destroyed.

This latter prediction would appear initially to be at odds with the observation of a pronounced mechanical anisotropy in moulded components immediately following fabrication. However, more recently, Folkes and Nazockdast[23] have carried out flow birefringence measurements on Kraton 102 (see Fig. 11) and the results show that for low shear rates, there is virtually no dependence of measured birefringence on shear rate and that the value of birefringence is close to that associated with the form birefringence of the single-crystal samples, discussed earlier. This suggests that the material is able to flow while maintaining its microphase morphology. At higher shear rates, the birefringence increases rapidly but after cessation of flow the measured birefringence recovers to the form birefringence value, in a time scale of the order of tens of seconds. This period is sufficiently brief to ensure that the recovery process is virtually complete during the hold-period in the injection-moulding cycle. In view of the fact that no further heat treatment of the injection-moulded part is required to develop the equilibrium microphase morphology, it seems likely that this is essentially retained over the whole range of shear rates. The transient increase in bire-

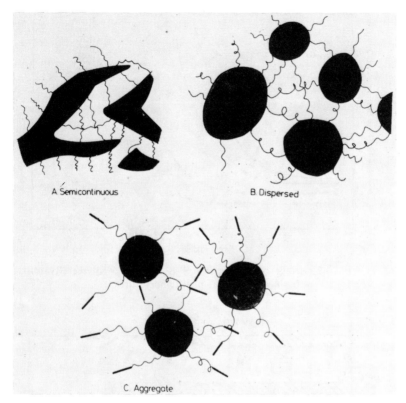

A. Semicontinuous   B. Dispersed

C. Aggregate

FIG. 10.   Schematic diagram showing the proposed microstructural changes
occurring during the melt flow of an S–B–S block copolymer.[26]

fringence at high shear rates is likely to be associated with the development and recovery of molecular orientation in the highly entangled
polybutadiene phase.

In order to interpret the origin of the polystyrene rod orientation
distribution in the plaque moulding studied by Folkes and Nazockdast[23]
and in other simple mouldings, it is necessary to use some of the ideas
advanced by Tadmor.[27] With reference to Fig. 12, it can be seen that on
entering the cavity, the fluid will initially be subjected to compressional
flow, leading to a divergence of the streamlines away from the gate.
Simultaneously, a solidified layer of polymer forms on the mould surface.
A flow front establishes itself, advancing forward by the flow of the
molten polymer through a channel defined by the boundaries of the

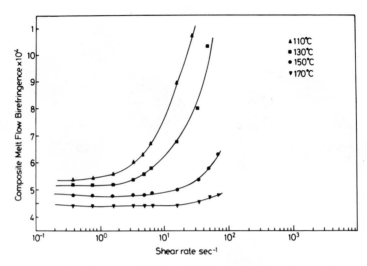

FIG. 11.   Flow birefringence data for the S–B–S block copolymer Kraton 102
obtained at various melt temperatures.[23]

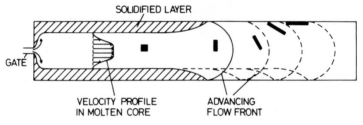

FIG. 12.   Schematic diagram of the mould-filling process showing the deformation of an initially square fluid element at successive positions of the advancing melt front.

solidified polymer — often referred to as the 'skin' layer. The velocity profile in the 'core' region will correspond to the non-isothermal flow of a non-Newtonian fluid. Further towards the flow front, the velocity field becomes considerably more involved. In a Lagrangian frame of reference (i.e. one in which the observer moves with the same velocity as the advancing flow front) the motion of the fluid is similar to that of a 'fountain', with fluid elements decelerating as they approach the flow front from the core and acquiring a radial component of velocity as they move towards the wall. In a laboratory frame of reference, the actual change in orientation of the fluid elements during this process is shown in Fig. 12. It is clear, therefore, that the elongational flow field at the flow front will be very effective in aligning the microphases along the axis of the moulding in the skin layer, as found experimentally. The orientation of the microphases in the core region will depend on the detailed shape of the velocity profile but their orientation will change progressively from the skin to the axis of the moulded component.

4.4. Microstructure and Properties of Block Copolymer–Homopolymer Blends

The foregoing discussion has shown that significant mechanical aniso-tropy can develop in components in S–B–S block copolymers. The anisotropy is little affected by processing conditions but significantly so by the geometry of the injection mould cavity. The anisotropy that is developed may be inappropriate for a particular application and, accord-ingly, an alternative method of microstructural modification is called for. The mechanical blending of two or more homopolymers to achieve new properties is of widespread interest and the potential of blending block copolymers with homopolymers to produce materials of commercial importance has been noted — see Chapter 1. In general, when an S–B–S block copolymer is blended with a commercial homopolymer, e.g. poly-styrene, a three-phase structure is produced. If the polystyrene is in the lesser proportion, then it will form a dispersed phase of roughly spherical particles in a matrix of the block copolymer, which itself is two-phase. In principle, this blend should produce a near isotropic material, owing to the disruption of the orientation in the block copolymer, and therefore provides additional control of the anisotropy of a moulded component. In the case where the block copolymer is in the lesser proportion, the blend manifests many of the characteristics of commercial high impact polystyrene.

The main features of block copolymer–homopolymer blends have been established through work on solvent-cast films. Kawai and his co-workers[28] and Skoulios et al.[29] have shown that when the molecular weight of the added homopolymer is less than or comparable to that of the corresponding block of the copolymer, solubilization of the homo-polymer into the microphases of the block copolymer takes place. In such cases, the dimensions of the microphases increase in relation to the amount of the added homopolymer and in some cases, the morphology of the microphase may also change. However, when the molecular weight of the added homopolymer is larger than that of the corresponding block copolymer, the block copolymer behaves as if it were largely incompatible with the corresponding homopolymer and a three-phase structure is produced. However, in such cases, the blocks in the interfacial region between the homopolymer and block copolymer are preferentially anchored in the homopolymer phase and thus help to form a stable interface. This is an important contributory factor to the success of these types of material as high impact polymers.

The surfactant properties of block copolymers when incorporated into homopolymers are well established. However, during melt processing of block copolymer–homopolymer blends, large stresses will be imposed on the two constituents and this will be additional to any interfacial interactions. The microstructure and properties of melt-processed blends of homopolymers have been studied extensively (see, for example, Ref. 30) and it has been shown that under appropriate conditions the dispersed phase can deform during processing to produce fibres. The microrheological interpretation of this has been reported by Han[31] and by Lyngaae-Jørgensen in Chapter 3 of this present book. The corresponding studies using a block copolymer–homopolymer blend are far less numerous. Nandra et al.[32] have reported the microstructure and mechanical properties of blends of Shell TR-4113 and polystyrene resulting from a process of milling followed by injection moulding. Figure 13, taken

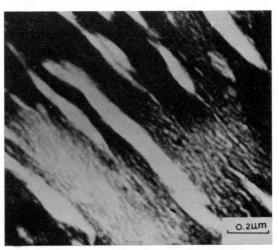

FIG. 13. Microstructure of injection mouldings of a blend of the S–B–S block copolymer TR-4113 with homopolymer polystyrene cut parallel to the major flow direction.[32]

from their work, shows that a three-phase structure exists in which the homopolymer polystyrene phase has deformed into ellipsoidal-shaped particles. Continuous polystyrene fibres were not observed. Although there are good reasons to expect improved stabilization of the drawing process occurring in the dispersed phase owing to the presence of the block copolymer,[33] it would seem likely that this cannot be achieved at the very high flow rates in injection moulding.

4.4.1. Screw Extrusion of S–B–S Copolymer–Polystyrene Blends

The mechanical properties of block copolymer–homopolymer blends in which the latter phase exists in fibril form would be of particular interest in the context of composite mechanics. Such a system has been produced unexpectedly as a result of the screw extrusion of blends of the S–B–S block copolymer, Kraton 102, with homopolymer polystyrene (Reip[34]). The block copolymer and a general purpose grade of polystyrene were compounded using a two-roll mill. Three concentrations of polystyrene were used, namely 10%, 30% and 50% by weight. After granulation, the blend was extruded using a Reifenhauser single-screw extruder fitted with a general-purpose screw. The die used in these studies is shown schematically in Fig. 14. A significant feature of this die is the presence of a spider consisting of a simple plate containing 18 circular holes. Although this is commonly used as a means of removing the rotational memory of the melt when in the screw, the melt will be subjected to elongational flow in its passage through each hole. It appears that this is important for the development of the particular microstructure to be described.

Figure 15 shows a transverse section cut from the extrudate of the 30% PS blend. This is viewed in the optical microscope using transmitted light. Rather suprisingly, a two-phase structure on the scale of $1 \mu m$ can be observed without resort to any of the usual methods used for revealing the dispersed phase in toughened plastics, e.g. etching techniques. The fact that the dispersed phase has dimensions $\sim 0 \cdot 5 - 1 \cdot 0 \mu m$ suggests that the contrast between the dispersed PS and the block copolymer matrix arises from scattering of the light at the PS–copolymer interface. Confirmation of the fibrillar nature of the homopolymer PS comes from scanning electron microscopy (SEM). Transverse sections of the extrudate were etched with a strong oxidizing acid, known to attack the rubbery matrix. Figure 16 shows the irregular break-up of the aligned structure. The fibrillar nature of the extrudate is also apparent from SEM observations of longitudinal fracture surfaces produced in three-point bending at low temperatures. A closer inspection of transverse

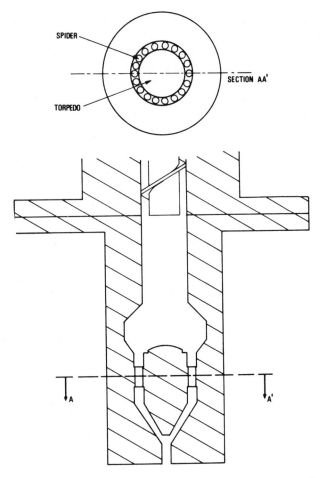

FIG. 14.   Diagram of the extrusion die used for the blend studies.[34]

sections in the optical microscope shows that the PS phase is arranged along well-defined layers. This is very surprising, since it implies a high degree of transverse ordering of the dispersed phase — an observation which to the best of our knowledge, has not featured in earlier rheological studies of two-phase fluids. Indeed it would be difficult to visualize a flow mechanism that would alone account for this structure. The layering is reminiscent of that observed by Dlugosz et al.[35] when a block copolymer exhibiting a lamellar morphology is subjected to an extrusion

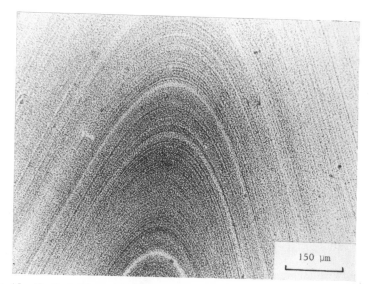

FIG. 15. Optical micrograph of a transverse section cut from a blend of the block copolymer Kraton 102 with 30% homopolymer polystyrene, following the processes of milling and screw extrusion.[34]

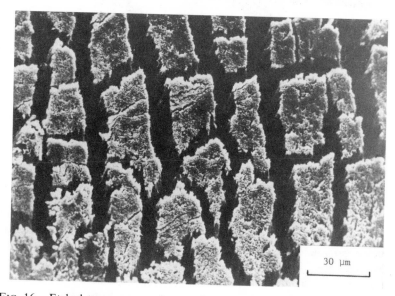

FIG. 16. Etched transverse section cut from a blend of Kraton 102 with 30% homopolymer polystyrene, following the processes of milling and screw extrusion.[34]

process. The layer periodicity, of course, is on a vastly different scale in the two cases. Nevertheless, the driving force required to develop transverse ordering in the blend could partly arise if the block copolymer matrix undergoes a phase transition to a lamellar morphology during the process of blending. Confirmation of this has been obtained by transmission electron microscopy using the Kato[36] osmium tetroxide staining technique — see Fig. 17. The phase transition is believed to occur by the

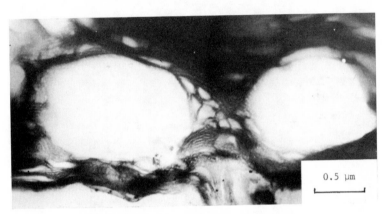

0.5 μm

FIG. 17. Transmission electron micrograph of a thin section cut transverse to the extrusion direction for a blend of Kraton 102 with 30% homopolymer polystyrene. The white areas are the unstained polystyrene phase.[34]

solubilization of a low molecular weight fraction of the homopolymer PS by the dispersed microphases in the block copolymer. The required volume fraction of low molecular weight material is either naturally present in the homopolymer or is additionally generated by chain scission during the compounding and screw extrusion processes.

A schematic diagram of the microstructure of the blend is shown in Fig. 18. The microstructure arises from a combination of factors, namely the process of phase transition in the block copolymer and the significant lowering of the surface energy of the dispersed homopolymer PS by the block copolymer. The latter aids the formation of the large fibrils by the process of elongational flow through the 'spider' holes — a mechanism more generally described by Lyngaae-Jørgensen elsewhere in this book. Finally, the fibrils are ordered by their association with the block copolymer.

It is interesting to note that the formation of this microstructure

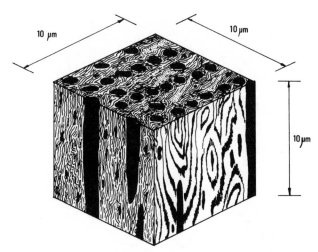

FIG. 18. Schematic diagram of the microstructure of the Kraton 102–30% homopolymer polystyrene blend. The large black areas represent the homopolymer polystyrene.[34]

depends on a delicate balance of the factors just described and that it is partly fortuitous that the screw extrusion process can generate this. For example, alterations to the volume fraction of added PS have a marked effect on the blend microstructure; for 10% PS, the blend shows no evidence of fibril formation (see Fig. 19), whereas for 50% PS almost perfect fibres are produced (see Fig. 20). Also, a more intensive mixing process as developed in a Transfermix compounding extruder, produces a blend that is transparent, i.e. the added homopolymer PS must be present as particles with dimensions much smaller than the wavelength of light. Fibrils are not produced in this case. Some fine particles of this type are also found in the 30% PS blend produced on the screw extruder.

The proposed microstructure of the 30% PS blend is consistent with swelling measurements made by Reip.[34] Small samples were swollen in hexane vapour, known to be a selective swelling agent for polybutadiene. The swelling of the sample was anisotropic as shown in Fig. 21. No dimensional changes were observed along the original direction of extrusion, i.e. the fibril direction. The swelling in the other directions was also anisotropic, it being greater perpendicular to the layer structure compared with that within a layer.

Block copolymer–homopolymer blends represent very complex materials from the standpoint of composite mechanics. The blend discussed

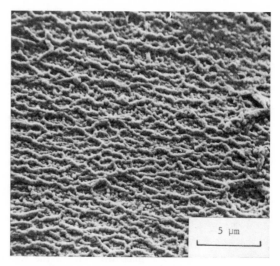

FIG. 19. Scanning electron micrograph of an etched longitudinal section cut from a blend of Kraton 102 with 10% homopolymer polystyrene, following the processes of milling and screw extrusion.[34]

FIG. 20. Scanning electron micrograph of an etched longitudinal section cut from a blend of Kraton 102 with 50% homopolymer polystyrene, following the processes of milling and screw extrusion.[34]

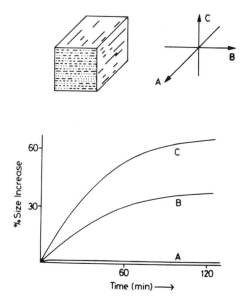

FIG. 21. Swelling anisotropy in the blend of Kraton 102 with 30% homo-polymer polystyrene. A represents the original direction of extrusion while B is tangential to the striations shown in Fig. 15.[34]

above has a matrix, which itself is a composite of polybutadiene and polystyrene, and a homopolymer polystyrene phase which is present as aligned fibrils and as a finely dispersed phase (< 5% vol.). It poses a major challenge, even in the prediction of the longitudinal Young's modulus. Small samples cut from the 30% PS blend exhibited a stress–strain curve as shown in Fig. 22. The slope of the initial part of the plot provides a measure of Young's modulus and was found to be $E_0 = 0.46$ GPa. A theoretical estimate of the longitudinal Young's modulus may be made as follows.

The Kraton 102 matrix contains 20% vol. of polystyrene. Since 30% vol. of polystyrene has been added to form the blend, the total proportion of polystyrene in the blend including that in the block copolymer will be $(0.2 \times 0.7) + 0.3 = 0.44$. After blending, a proportion of the added homopolymer polystyrene is occluded in the polystyrene microphases to effect a phase transition from a cylindrical to lamellar morphology. The residue takes the form of the aligned fibrils or as a fine dispersion of approximately spherical particles (< 5% vol.). An estimate of the volume

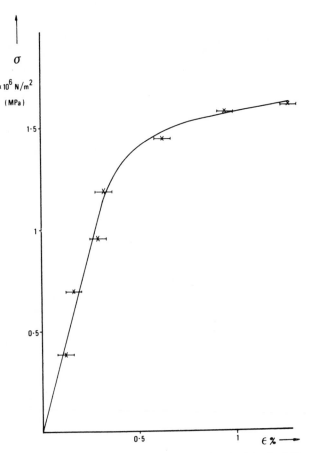

FIG. 22. Typical longitudinal stress–strain curve for the blend of Kraton 102 with 30% homopolymer polystyrene, following the processes of milling and screw extrusion.[34]

fraction of fibrils in the blend may be obtained from measurements of fibril diameter and spacing on scanning electron micrographs and is found to be 10%. Hence, there is 30–35% of polystyrene in the block copolymer after blending, which is just sufficient to promote a phase transition to a lamellar morphology.

Now, to a first approximation, the longitudinal Young's modulus of the blend will be given by

$$E_0 = V_f E_f + V_B E_B$$

where E_f and E_B are the Young's moduli of the fibrils and block copolymer, respectively, and V_f and V_B are the volume fractions of the fibrils and block copolymer, respectively.

This assumes that the aspect ratio (length/diameter ratio) of the fibrils is sufficiently large so that a Voigt average is valid. An upper and lower bound to the modulus of the blend may be obtained by assuming either that the lamellae in the block copolymer are fully aligned along the extrusion direction or that they are randomly oriented in three dimensions. In practice, they are partially oriented owing to the competing effects of flow and the disrupting influence of the small proportion of finely dispersed homopolymer polystyrene.

Upper bound Since the modulus of polystyrene greatly exceeds that of the polybutadiene, $E_B \approx V E_f$, where V is the volume fraction of polystyrene in the copolymer. Hence,

$$V_f = 0{\cdot}1 \qquad V_B = 0{\cdot}9 \qquad V = 0{\cdot}34$$
$$E_f = 2{\cdot}0 \, \text{GPa} \qquad E_B = 0{\cdot}68 \, \text{GPa}$$

Therefore

$$E_0 = 0{\cdot}1 \times 2{\cdot}0 + 0{\cdot}9 \times 0{\cdot}68 = 0{\cdot}81 \, \text{GPa}$$

Lower bound In this case the block copolymer is randomly oriented, giving $E_B \sim (1/6) \times 0{\cdot}68 = 0{\cdot}11 \, \text{GPa}$

$$E_0 = 0{\cdot}1 \times 2{\cdot}0 + 0{\cdot}9 \times 0{\cdot}11 = 0{\cdot}3 \, \text{GPa}$$

The experimental value of E_0 correctly lies between the upper and lower bounds.

Transmission electron microscopy of osmium tetroxide stained samples of the blend[34] has confirmed that the lamellae in the block copolymer are partially aligned. However, further work will be needed to characterize fully the orientation distribution function (using, for example, low angle X-ray scattering) so that a more precise value for the longitudinal Young's modulus can be predicted using the form of analysis presented earlier in this chapter.

REFERENCES

1. Hearmon, R. F. S. (1961). *Introduction to Applied Anisotropic Elasticity*, Oxford University Press, Oxford.
2. Arridge, R. G. C. and Barham, P. J. (1982). *Adv. Poly. Sci.*, **46**, 69.

3. Arridge, R. G. C. (1985). *An Introduction to Polymer Mechanics*, Taylor and Francis, London, (in press).
4. Van Fo Fy, G. and Savin, G. (1965). *Mech. Polim.*, **1**, 151.
5. Walpole, L. J. (1969). *J. Mech. Phys. Solids*, **17**, 235.
6. Timoshenko, S. P. and Woinowsky-Krieger, S. (1959). *Theory of Plates and Shells*, (2nd edn) McGraw-Hill, New York.
7. Tsai, S. W. (1964). *Structural behaviour of composite materials*, NASA CR-71.
8. Keller, A., Pedemonte, E. and Willmouth, F. M. (1970). *Nature*, **225**, 538.
9. Arridge, R. G. C. and Folkes, M. J. (1972). *J. Phys. D: Appl. Phys.*, **5**, 344.
10. *idem.* (1975), *ibid.*, **8**, 1058.
11. Filon, L. N. G. (1902). *Phil. Trans. Roy. Soc.*, **A198**, 147.
12. Toupin, R. A. (1965). *Arch. Rat. Mech. Anal.*, **18**. 83.
13. Horgan, C. O. (1972). *J. Elast.*, **2**, 335.
14. Arridge, R. G. C., Barham, P. J., Farrell, C. J. and Keller, A. (1976). *J. Mater. Sci.*, **11**, 788.
15. Aggarwal, S. L. (1976). *Polymer*, **17**, 938.
16. Folkes, M. J., Keller, A. and Scalisi, F. P. (1973). *Kolloid–Z. u. Z. Polymere*, **251**, 1.
17. Charrier, J. M. and Ranchoux, R. J. P. (1971). *Polym. Eng. Sci.*, **11**(5), 381.
18. Beecher, J. F., Marker, L., Bradford, R. D. and Aggarwal, S. L. (1969). *J. Polym. Sci.*, **26C**, 117.
19. Morton, M., McGrath, J. E. and Juliano, P. C. (1969). *J. Polym. Sci.*, **26C**, 99.
20. Lewis, P. R. and Price, C. (1971). *Polymer*, **12**, 258.
21. Harpell, G. A. and Wilkes, C. E. (1969). ACS Division of Polymer Chemistry paper presented at New York meeting, Sept. 8–12.
22. Mieras, H. J. M. A. and Wilson, E. A. (1973). *JIRI*, **7**(2), 72.
23. Folkes, M. J. and Nazockdast, H. (1985). To be published.
24. Folkes, M. J. and Keller, A. (1971). *Polymer*, **12**, 222.
25. Puppo, A. H. and Evensen, H. A. (1970). *J. Comp. Mat.*, **4**, 204.
26. Arnold, K. R. and Meier, D. J. (1970). *J. Appl. Polym. Sci.*, **14**, 427.
27. Tadmor, Z. (1974). *J. Appl. Polym. Sci.*, **18**, 1753.
28. Inoue, T., Soen, T., Hashimoto, T. and Kawai, H. (1970). In: *Block Polymers*, (Ed. S. L. Aggarwal), Plenum Press, New York, pp. 53–79.
29. Skoulios, A., Helffer, P., Gallot, Y. and Selb, J. (1971). *Makromol. Chem.*, **148**, 305.
30. Manson, J. A. and Sperling, L. H. (1976). *Polymer Blends and Composites*, Plenum Press, New York.
31. Han, C. D. (1981). *Multiphase Flow in Polymer Processing*, Academic Press, New York.
32. Nandra, D. S., Hemsley, D. A. and Birley, A. W. (Feb. 1979). Plast. Rubb.: Mat. Appl., p. 38.
33. Meier, D. J. (1981). Private communication.
34. Reip, P. W. (1983). Ph.D. Thesis, Brunel University, Uxbridge, Middlesex, England.
35. Dlugosz, J., Folkes, M. J. and Keller, A. (1973). *J. Polym. Sci.*, **11C**, 929.
36. Kato, K. (1966). *J. Polym. Sci.*, **4B**, 35.

APPENDIX

The relations between the two-suffix notation and the tensor notation are, for stiffnesses,

$$C_{aabb} = C_{ab}(a,b = 1,2,3)$$
$$C_{aabc} = C_{ad}(a,b,c = 1,2,3; \ d = 4,5,6)$$
$$C_{abcd} = C_{ef}(a,b,c,d = 1,2,3; \ e,f = 4,5,6)$$

and the pairs of suffixes 11, 22, 33, 23 or 32, 31(13), 12(21) become, respectively, 1, 2, 3, 4, 5, 6. Thus $C_{1111} \rightarrow C_{11}$, but $C_{1212} \rightarrow C_{66}$.

For the compliances, however, we have

$$S_{aabb} = S_{ab}$$
$$2S_{aabc} = S_{ad}$$
$$4S_{abcd} = S_{ef}$$

Stresses σ_{ij} and strains e_{ij} in the tensor notation transform as follows: $\sigma_{11}, \sigma_{22}, \sigma_{33} \rightarrow \sigma_1, \sigma_2, \sigma_3$ but the shear stresses $\sigma_{23}, \sigma_{31}, \sigma_{12} \rightarrow \sigma_4, \sigma_5, \sigma_6$. Strains $e_{11}, e_{22}, e_{33} \rightarrow e_1, e_2, e_3$, but the shears require a factor 2 in the transformation so that $2e_{23}, 2e_{31}, 2e_{12} \rightarrow e_4, e_5, e_6$.

If the direction cosines of the axes (1', 2', 3') with respect to the set (1, 2, 3) are $l_{i'j}$ then the tensor C_{ijkl} in the (1, 2, 3) axes becomes

$$C_{i'j'k'l'} = l_{i'i}l_{j'j}l_{k'k}l_{l'l}C_{ijkl}$$

with automatic summation over the indices i,j,k,l from 1 to 3. $C_{i'j'k'l'}$ has, therefore, 81 components.

The direction cosines for the transformation of axes in Fig. 2 are

$$l_{1'1} = \cos\psi \cos\theta \cos\phi - \sin\psi \sin\phi$$
$$l_{1'2} = -\cos\psi \sin\phi - \sin\psi \cos\theta \cos\phi$$
$$l_{1'3} = \sin\theta \cos\phi$$
$$l_{2'1} = \cos\psi \cos\theta \sin\phi + \sin\psi \cos\phi$$
$$l_{2'2} = -\sin\psi \cos\theta \sin\phi + \cos\psi \cos\phi$$
$$l_{2'3} = \sin\theta \sin\phi$$
$$l_{3'1} = -\cos\psi \sin\theta$$
$$l_{3'2} = \sin\psi \sin\theta$$
$$l_{3'3} = \cos\theta$$

(Note: ψ measures the rotation of the 1-axis out of the plane defined by 3' and 3).

Then the mean value of $C_{i'j'k'l'}$ is given by

$$C_{i'j'k'l'} = \int\int\int l_{i'i}l_{j'j}l_{k'k}l_{l'l}C_{ijkl}\Phi(\theta,\phi,\psi)\sin\theta\mathrm{d}\theta\mathrm{d}\,\phi\mathrm{d}\psi$$

and an analogous expression for $S_{i'j'k'l'}$.

The two-suffix symbols may then be derived from the tensors given above, using the symmetry of the problem.

CHAPTER 5

Segmented Copolymers with Emphasis on Segmented Polyurethanes

SAADEDDINE ABOUZAHR and GARTH L. WILKES

Polymer Materials and Interfaces Laboratory, Department of Chemical Engineering, Virginia Polytechnic Institute and State University, Blacksburg, USA

1. INTRODUCTION

The aim of this chapter is to provide a general overview of the structure/property behaviour of segmented copolymers, with the principal emphasis being on segmented elastomers. While the authors have emphasized the segmented urethanes, information regarding the segmented polyether/polyester materials, as well as some information on the segmented urethane–urea materials, is also provided. Where possible, contrast is made with the 'long block' ABA block copolymers discussed in several of the other chapters within this book. While the authors provide several comments concerning specific work of other researchers in the literature, caution should be taken with regard to readily applying the structure–property behaviour from one system to that of another of varied chemistry. Rather, it is hoped that the information provided by this chapter will induce some understanding of the general structure/property aspects that arise from the wide diversification of chemistry and processing variables that may influence the final behaviour of these materials.

Thermoplastic nonionic elastomeric copolymers are macromolecular systems that exhibit several aspects of rubber-like elasticity but are not covalently crosslinked.[1] The unique feature of these materials is that when viewed at the *use* temperature, the polymer chains are composed of two chemically incompatible segments or blocks. One of these segments, generally known as the hard block, is glassy or crystalline owing to the fact that it is below its T_g or T_m at the use temperature. These hard blocks associate to form small and distinct morphological domains that serve as physical crosslinks and reinforcement sites, hence limiting flow of the other phase made up of the 'soft segments' or rubbery blocks which are above their T_g at the use temperature. Since the hard domains soften at their T_g or T_m, a thermally reversible processable elastomeric pseudo network system results as long as the two segments are chemically incompatible, which in turn promotes the thermodynamic driving force for phase separation.

Being very different from vulcanized rubber in a chemical sense, thermoplastic elastomeric systems can be extruded and solution-processed, as well as recycled, owing to the fact that no covalent cross-links exist. In general, it is the thermal span between the soft and hard segment T_g that dictates the range of service temperature for their use as 'elastomers'. As will be discussed shortly, other factors besides temperature strongly influence mechanical properties.

Segmented polyurethanes and several other segmented copolymers differ from the diene type of block copolymers in the following respects. First, the number of hard and soft segments in a sequential system is much greater than the typical diene block copolymer. Secondly, the molecular weight of both segments is typically low $(400\text{--}6000\,\mathrm{g\,mol}^{-1})$ relative to values of $10\,000$ to $100\,000\,\mathrm{g\,mol}^{-1}$ for the ABA block copolymers. The segmented systems often (though not necessarily) have inter- and intra-molecular hydrogen bonding, whereas typical ABA block copolymers of the diene type generally do not possess this type of secondary bonding. Owing to their short length and stiffness (at least for several hard segments) the hard segments tend, therefore, to behave more 'rod-like', in contrast to a long polystyrene coiled glassy block in an S–B–S copolymer. Likewise, the 'shortness' of the soft segments tends to limit the degree of extensibility. One outcome that arises from the short chain nature of the segments as well as the stiffness of the hard segment, is that these systems have provided several obstacles to those theoreticians who might attempt phase separation calculations as has been done in the non-polar long block copolymers of the diene type. In addition, owing to the segmented nature, there are many more chemical junctions between the unlike components, in contrast to the long block systems. This increase in junction number has important consequences in terms of entropic restrictions, thereby affecting the thermodynamic considerations of the phase separation process. The possible inapplicability to utilize Gaussian chain considerations for these short segments can also hinder theoretical developments. Figure 1 schematically illustrates aspects of the two-phase nature in the segmented systems and contrasts this structure to that of the styrene/butadiene block copolymers.

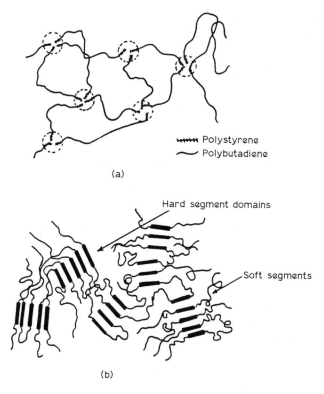

Polystyrene
Polybutadiene

(a)

Hard segment domains

Soft segments

(b)

FIG. 1. Schematic representation of long block S–B–S and short block or segmented urethane domain morphologies.

2. CHEMISTRY OF SEGMENTED COPOLYMERS

Most segmented polyurethanes are formed in such a way that the chemical reaction employs both monomer and preformed functionally terminated oligomers.[2] The majority of polyurethanes commercially available are based upon low molecular weight $(600–6000\,\mathrm{g\,mol^{-1}})$ polyester or polyether soft segments that are terminated with hydroxyl groups. Recently some hydroxyl-terminated hydrocarbon polymers, such as polybutadiene,[3-6] polyisobutylene,[7] or polysiloxane,[8] have been utilized. The other starting materials or intermediates consist of di- or polyfunctional isocyanates and with most systems, low molecular weight polyfunctional alcohols, amines, or acids. Several chemical reactions may be involved in the polymerization process. The diisocyanate can react with either the soft segment or the chain extender, as shown below:[9]

$$RNCO + R'OH \longrightarrow RN-\underset{\underset{O}{\|}}{C}-O-R'$$

urethane linkage

$$RNCO + R'NH \longrightarrow \underset{}{R}N-\underset{\underset{O}{\|}}{C}-\underset{}{N}-R'$$

with H above the first N and H above the second N

urea linkage

$$RNCO + R'COOH \longrightarrow R-\underset{\underset{H}{|}}{N}-\underset{\underset{O}{\|}}{C}-R' + CO_2$$

amide linkage

In practice, a homologous mixture of short chain end-capped iso-cyanate terminated oligomer is produced in these reactions. These macro-diisocyanates can then be reacted with more diol to produce a linear polymer chain; or chain extension may be carried on by reaction of the macro-diisocyanate with a low molecular weight diol or diamine — the latter leading to the formation of urethane–urea linkages. The preceding scheme is a two-step process but it is also possible to use a one-step process by varying the stoichiometry and adding the chain extender moiety in the first step. Another possible modification is to use water as a chain extender which results in the evolution of carbon dioxide which in turn produces a urea linkage (NHCONH) such as that shown below:

$$RNCO + H_2O \longrightarrow RNHCOOH$$

$$\downarrow$$

$$RNH_2 + CO_2$$

$$\downarrow$$

$$RNHCONHR$$

urea

While the above reactions are rather straightforward there is always the possibility of competitive side reactions. Close attention must be given to reaction rates, order of addition of ingredients and catalysis of certain reactions in preference to others. For example, the isocyanate is capable of reacting with the active hydrogen on the urethane or urea group to give branching or crosslinking as indicated below:

$$RNCO + R'NHCOOR' \longrightarrow \begin{matrix} R'NCOOR'' \\ | \\ CONHR \end{matrix}$$

allophanic ester formation

$$RNCO + R'NHCONHR' \longrightarrow \begin{matrix} R'NCONHR'' \\ | \\ CONHR \end{matrix}$$

biuret formation

The RIM (reaction injection moulding) materials based on urethanes are also segmented in nature. Here the liquid polyol, chain extender, catalyst, and diisocyanate are directly mixed in a static mixing head, after which the mixture flows into the hot mould where the polymerization reaction transpires quickly, leaving the final bulk polymer which can be removed from the mould as a finished part. Owing to the fact that RIM systems may be used for structural applications, as in auto body components, they may need to be of higher hardness and rigidity, in addition to being amenable to fast moulding times. Hence the chemistry differs somewhat from the linear thermoplastic urethane elastomers, in that the polyol may be of higher molecular weight and may contain trifunctional species which promote a covalent network. Polyfunctional chain extenders and isocyanates may also be used to achieve the same purpose, although in certain cases these species may limit the rigidity of the hard phase because of difficulty in packing caused by the architecture of a trifunctional moiety. High modulus and low heat sag characteristics are often promoted by utilizing a symmetric hard segment that will allow the formation of crystalline domains. Rigid, aromatic hard segments can also achieve the same purpose even if they promote the formation of totally amorphous domains. Variation in the rigidity of the polyol, as well as functionality and symmetry, are all features which promote the development of good physical properties and the desirable impact characteristics of RIM systems. Since the processability of RIM materials is as important as their physical properties, crystallizable polyurethanes may not always be adequate since they are more sensitive to fluctuations in processing conditions. Amorphous systems, on the other hand, have lower physical properties but processibility could be significantly better than their crystalline counterparts. The technology of the RIM process has been discussed in detail elsewhere[10] and will not be given here.

One advantage of the basic polymerization schemes discussed above is that they are typically economical and allow the possibility of bulk as well as solution polymerization. Their polymerization, however, does suffer from the drawback that the molecular characteristics of the second block or hard segment cannot be easily determined separately from the final block copolymer, as is often the case in the diene type of block copolymers.[2] Furthermore, in —(A—B)— segmented copolymers, the preformed oligomers generally enter the block copolymer in a statistical rather than a perfect alternating fashion, as may be the case when the oligomeric end-groups are functionally identical to that of one of the step-growth monomers (chain extender).[2] However, the hard segment polydispersity of the segmented copolymer is a function of the relative reactivities of the oligomer, chain extender and isocyanate components.

The development of urethane elastomers began and grew around covalently crosslinked urethane systems, in keeping with the classical methods of producing elastomeric materials. In addition to biuret and allophanate crosslinks, crosslinking of these systems can also be promoted by the addition of organic peroxides.[11a,b] This generally involves hydrogen abstraction and the generation of free radicals so that carbon-to-carbon covalent crosslinks are induced. Furthermore, sulphur-accelerated curing systems can also be utilized to crosslink special structure polyurethane millable gums.[12] This is brought about by building occasional special curing sites into the polymer chain which favourably respond to the free radical cure mechanism. Such active structures are the ethylenically unsaturated side groups that can be attached to a urethane chain.[13] These groups can be readily incorporated via ethylenically unsaturated glycols, e.g. glyceryl-α-allylether or trimethylolpropane monoallylether as a chain extender [12-14] or macroglycol polymerization initiator.[14] For further details regarding crosslinking reactions of polyurethanes the reader is referred to Refs 11, 15 and 16. Additional studies displaying variation in polymerization of segmented urethanes or urethane–ureas, as well as variation in properties with chemistry, are described within Refs 17–28.

In contrast to the segmented urethanes, segmented polyester–ether copolymers are generally prepared by the melt transesterification of an alkyl terephthalate, a chain extender such as 1,4-butanediol, and poly(alkoxides) ethers[29-34] — an example being shown below where φ represents an aromatic moiety.

$$HO-\underset{\substack{\| \\ O}}{C}-\varphi-\underset{\substack{\| \\ O}}{C}-OH + HO-(CH_2)_4OH + HO-(R'-O)_xH$$

segmented copolyester polyether

where the hard segment is given by:

$$-(\underset{\substack{\| \\ O}}{C}-\varphi-\underset{\substack{\| \\ O}}{C}-O-(CH_2)_4O)-$$

and the soft segment by:

$$-(\underset{\substack{\| \\ O}}{C}-\varphi-\underset{\substack{\| \\ O}}{C}-O-(R'-O)_x)-$$

The overall structure of the polymer is that of hard and soft segments that are randomly arranged along the polymer chain. The hard segment is typically symmetric and therefore capable of crystallization leading to the formation of partially crystalline hard domains, the morphological texture of which is very much dependent on the preparation process. Schematically, this random arrangement could be described as:

$$\underline{T-B-T-B-T-B-T}-C$$

hard   soft

where example moieties are:

(B) $-(O-(CH_2)_4-O)-$;   (C) $-(O-(CH_2)_4-O)_{10}$

(T) $-O-(CH_2)_4-O-\underset{\substack{\| \\ O}}{C}-\varphi-\underset{\substack{\| \\ O}}{C}-$

The soft segment component usually has a value of M_n of about 1000–2000 and the hard segment M_n ranges between 600–3000 in copolymers of technical interest. Total molecular weight (weight average) may be as high as 200 000 to 400 000. If the hard segment length is determined by the ratio of the polyether to 1,4-butanediol, and assuming that the normal kinetics of polyester formation can be applied, both a hard and soft segment length distribution will result and can be identified as the sequence length distribution of T–C and T–B units.[33,35]

It has been demonstrated that the type of polyether glycol utilized has a pronounced effect on the synthesis of the polyether–ester copolymers.[30,34] This is evident from the wide range of inherent viscosities of these copolymers, the viscosities being influenced in turn by the soft segment type. High inherent viscosities have been obtained when poly(tetramethylene oxide) glycol (PTMO) and polyethylene oxide glycol (PEO) are employed, while low inherent viscosities will result when polypropylene oxide glycol (PPO) and ethylene oxide capped polypropylene oxide glycol (PEO–PPO) are used.[34] This behaviour has been attributed to three factors:[30,34] (1) At any given inherent viscosity, the melt viscosity of the PPO-based copolymer is significantly higher than that of the analogous PTMO- or PEO-based systems. Consequently the copolymer formation which is controlled by the diffusion-limited rate of removal of the 1,4-butanediol is retarded. (2) The rate of formation of the segmented copolymers based on PPO is decreased by the presence of a secondary hydroxyl end-group on the PPO monomer relative to the more reactive primary hydroxyl groups of the PTMO and PEO monomers. (3) The rate of thermal degradation of PPO-based copolymer is faster than that of those where the soft segment is PTMO or PEO. In addition the temperature at which these copolymers are prepared promotes the degradation of polypropylene oxide and reduces the molecular weight of the system. It has been pointed out[34] that substituting PEO-terminated PPO for PPO is of only marginal benefit in overcoming the problem of low molecular weight.

3. STRUCTURE/PROPERTY RELATIONSHIPS OF SEGMENTED POLYURETHANES

Several excellent studies, e.g. molecular weight property relationships,[36,37] permeability,[38] and mechanical behaviour,[36,37,39–41] have been published on the effect of diol, diisocyanate, and polyol types. A number of useful papers describing the processability and properties of polyurethanes can also be cited.[42-50] An attempt to illustrate some of the effects of these variables on properties is now given.

3.1. Effect of Polyol or Soft Segment Type

The flexible (soft) segments in polyurethane elastomers greatly influence the elastic nature of the material and significantly contribute to its low temperature properties and extensibility. Therefore the parameter of soft segment T_g is highly important. Furthermore crystallinity, if any, the melting point, and the possible ability to crystallize with strain, will also certainly influence the ultimate mechanical properties. As indicated earlier, hydroxy-terminated aliphatic polyesters and polyethers are the most common materials used to form the soft segments, although amine-terminated systems have also been reported — the latter will develop urea–urethane linkages when chain extended with a diol moeity.[51]

Polyesters generally give urethane elastomers a higher level of physical properties than the corresponding polyether-based materials, owing to the higher interchain attractive forces.[52-54] However, polyether-based polyurethanes have substantially better hydrolysis resistance — a feature reflecting the superior hydrolytic stability of ether groups over ester groups. Intermediate degrees of hydrolytic stability are usually achieved by utilizing low ester content polyesters[20] or polycarbonate acetals as soft segments.[20] In addition, urethanes based on polyethers display better low temperature properties than their ester counterparts. This is attributed to the substantially lower glass transition temperature of the polyether soft segments as compared with the polyesters possessing comparable numbers of methylene units and of comparable molecular weight. In both segment types, crystallinity is enhanced by regularity, and strain-induced crystallization can improve tear resistance and tensile strength[55-57] and at the same time increase the hysteresis characteristics of the polymer.[58] It is well known that conventionally crosslinked elastomers, such as natural rubber, have high tensile strength for similar reasons. However, natural rubber shows good recovery because its crystallinity is totally or largely lost when the deformation is removed — this may not always be the case for crystallizable soft segments.

The structure of the soft segment backbone is also a very important factor in controlling the properties of polyurethane elastomers. Table 1 lists some of the asymmetric soft segments which possess side groups along the backbone; these generally give significantly poorer mechanical properties than those from more symmetric linear systems.[17] This is largely owing to the side groups restricting strain-induced crystallization

TABLE 1

SOME GENERAL EFFECTS OF POLYOL SOFT SEGMENT STRUCTURE ON THE PHYSICAL PROPERTIES OF POLYURETHANE ELASTOMERS[a]

Polymer	Polyol/molecular weight	DSV[b]	Shore hardness	Tensile strength (psi)	Elongation (%)	300% modulus (psi)	T[c] (°C)
A	Poly(ethylene adipate)glycol/980	0·824	86 (A)	7 400	655	900	136
B	Poly(tetramethylene adipate)glycol/989	0·904	88 (A)	7 800	530	1 300	160
C	Poly(hexamethylene adipate)glycol/986	1·058	82 (A)	8 600	560	1 200	147
D	Poly(1,4-cyclohexyldimethylene adipate)glycol/1190	0·697	60 (D)	5 600	355	4 800	142
E	Poly(tetramethylene glycol)/974	0·935	90 (A)	5 300	725	1 000	130
F	Poly(propylene glycol)/1005	0·874	76 (A)	4 200	800	640	146

[a]Utilizing MDI and butanediol as the hard segment in all cases. Taken from Ref. 17.
[b]Dilute solution viscosity.
[c]Approximate processing temperature determined by dynamic extrusion rheometry.

of the soft segment. On the other hand, urethanes with polyester soft segments containing bulky or rigid (e.g. aromatic) moieties often make tougher materials than those based on linear aliphatic blocks, as a consequence of the reduced flexibility of the soft segments. As expected, however, these same systems undergo a stiffening at low temperatures while more flexible linear soft segments retain their elastomeric properties to even lower temperatures.

A related explanation to account for low temperature properties is the correlation of the mechanical properties with the glass transition temperature of the soft segment, i.e. stiff chains containing rigid aromatic moieties that display the shortest separation of ester groups will possess the highest T_g's and consequently tend to produce tougher materials that harden at higher temperatures than those with more flexible aliphatic chains.[28] As stated above, the capability of the soft segment to crystallize adds another dimension.

As expected, variation of the soft segment structure will also alter the hysteresis and permanent set characteristics of urethane elastomers. In general, soft segments that possess rigid ring structure or bulky side groups display higher permanent set and hysteresis values.[9,28] On the other hand, symmetric soft segments that crystallize with deformation may also display high hysteresis properties if residual crystallization is retained by the soft phase after the stress is removed.[59] In this latter case, hysteresis properties can often be significantly reduced by disrupting the symmetry of the soft segment. This can be done by using trifunctional soft segments that interfere with the soft segment packing and hinder its strain-induced crystallization. Another variation is to use copolyesters or copolyethers which possess structural irregularity in a manner similar to hindering soft segment packing.[59] It should be pointed out, however, that soft segment crystallization is believed beneficial for improving tear strength properties at high elongations. As is always the case, a compromise must be made with respect to a balance of suitable physical properties.

As already indicated, the molecular weight of the soft segment in polyurethanes typically ranges from 600 to about 6000 g mol^{-1}. The optimum molecular weight, however, depends upon features such as hard and soft segment types, molar ratio of the components and, in general, on the specific formulation of the desired polyurethane system. For example, urethanes employing diphenyl methane diisocyanate (MDI), chain-extended with 1,4-butanediol as hard segments and polytetramethylene adipate as the soft phase, and in the molar ratio

(MDI/butanediol/soft segment) 1·7/1·0/1·7, have been stated to show 'optimum properties' when the soft segment molecular weight is between 600 and 1200 g mol^{-1}.[60] On the other hand, elastomers derived from polyethylene adipate and 1,5-naphthylene diisocyanate, chain-extended with water, show desirable properties when the polyester molecular weight is 2000 g mol^{-1}.[61] It should be recognized, however, that 'optimal properties' may convey a different meaning to those considering the applications of these materials and therefore caution must be taken when intrepreting the literature with respect to 'optimal properties'.

For linear and symmetric soft segments, the tendency for strain-induced crystallization is promoted, as the soft segment molecular weight is increased — other factors being constant. This may well result in better tear strength but will also tend to reduce hysteresis characteristics of the polymer, as stated earlier. At low soft segment molecular weight, the extent of phase separation is generally reduced, thereby also promoting poor hysteresis behaviour of the polymer. As a result, an optimum segment molecular weight exists at which the polymer will display its lowest hysteresis properties.[62]

Most recently, amine-terminated polyether polyols have been reported to be superior to their hydroxyl-terminated counterpart in RIM application.[51] It is entirely possible that the higher reactivity of the amine results in a selectively higher polyol–isocyanate coupling, which in turn results in high molecular weight build-up in these systems. This same higher reactivity might result in non-wettability of the mould surface during injection which may account for the better self-release capability of these materials. In addition, the basic nature of amine-terminated polyol allows the use of acidic-type mould release agents, which will normally degrade the strength of hydroxyl terminated systems.

In addition to the polyether or polyester soft segment types, hydroxyl-terminated butadiene has been used in some studies.[3-6] The latter component, of course, is of lower polarity and, therefore, less compatible in general with the hard segment phase. It has been difficult to date to delineate all the effects of changing from an ether or ester segment to that of the butadiene. This is owing to the fact that the functionality of the hydroxyl-terminated butadienes is generally slightly higher, and, therefore, has led to some crosslinking effects which have complicated the matters of structure/property behaviour. It is safe to say, however, that because of the more hydrophobic nature of this soft segment, better phase separation might be expected at the time of solidification from the melt.

In addition, the recent work of Yilgor *et al.*[63] and Tyagi *et al.*[64] has provided segment urea systems from siloxane oligomers that were terminated with amino propyl groups. These workers have developed segmented ureas by reacting the amino propyl-terminated dimethyl siloxane oligomers directly with MDI based on a stoichiometric balance. The resulting materials which ranged from the order of 6 wt% MDI (hard segment) to 20 wt% MDI showed elastomeric properties at room temperature and were far superior to unfilled high molecular weight dimethylsiloxane rubber. An example of the chemistry is illustrated below while Fig. 2 depicts some comparisons of the stress/strain behaviour with those of siloxane rubber and two filled siloxane elastomers.

$$H_2N-(CH_2)_3 \left[\begin{array}{c} CH_3 \\ | \\ Si \\ | \\ CH_3 \end{array} -O \right]_{<} \begin{array}{c} CH_3 \\ | \\ Si \\ | \\ CH_3 \end{array} -(CH_2)_3-NH_2$$

$$+ OCN-\varphi-CH_2-\varphi-NCO \longrightarrow$$

$n = 10$
amino propyl-terminated siloxane   MDI

$$\left[\begin{array}{c} O \\ \| \\ C \\ | \\ H \end{array} -N-\varphi-CH_2-\varphi-N- \begin{array}{c} H \\ | \\ C \\ \| \\ O \end{array} -N-(CH_2)_3 \left[\begin{array}{c} CH_3 \\ | \\ Si \\ | \\ CH_3 \end{array} -O \right]_{<} \begin{array}{c} CH_3 \\ | \\ Si \\ | \\ CH_3 \end{array} -(CH_2)_3-N \begin{array}{c} H \\ | \\ \end{array} \right]_x$$

segmented siloxane urea

It is apparent, based on these data as well as other data given by these workers, that the siloxane soft segmented urea-linked systems display a microphase texture that arises at very low weight percentages of MDI incorporation. Since no chain extender was utilized in these studies, it is also clear then that the hard segment itself is made up only of a single MDI unit and it will require structuring of these units to promote domain formation, which clearly must promote a bundle-like texture to the individual domain regions. At this stage there is no additional information regarding the detailed structure/property behaviour of these

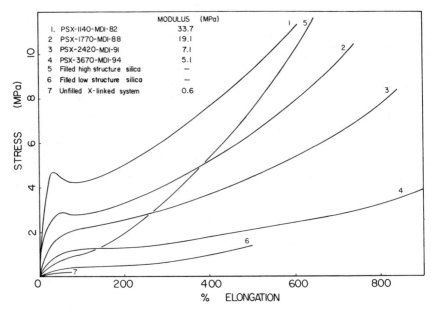

FIG. 2.   Stress vs. percentage elongation for several siloxane urea segmented copolymers as a function of siloxane oligomer molecular weight. Measurements were made at 25°C. Also shown for comparison is the stress/strain behaviour for two silica-filled dimethyl siloxane elastomers and an unfilled crosslinked siloxane elastomer.[64]

interesting materials but it is clear that based on the initial data and some of the unique features of siloxane itself, with time, these materials may find unique applications.

In summary, the type of soft segment, its symmetry, glass transition temperature and molecular weight, will affect the morphology and consequently the mechanical properties of urethane or urethane–urea elastomers. Symmetry enhances crystallization and contributes to higher modulus and higher tear strength, however, high hysteresis and permanent set may be noted depending on some of the other factors discussed above. Lower hysteresis and permanent set values are obtained with materials where soft–hard segment incompatibility is high (polyethers show better hysteresis properties than polyesters, other factors being equal) and by disrupting the regularity of the soft segment so that crystallization is impaired. Owing to the influence of the many 'segment' variables stated above, further consideration of the hard and soft segments is now given.

4. HARD SEGMENT TYPE

The hard segments in polyurethanes typically consist of an isocyanate and chain-extender glycol or amine components. Several types of diisocyanate (aromatic, aliphatic, cycloaliphatic) and many different chain extenders (open-chain aliphatic, cycloaliphatic, aromatic aliphatic) can be used. The proper choice of the hard segment type depends upon factors such as the desired polymer mechanical properties, upper service temperature, environmental resistance, solubility characteristics and, of course, economics.

4.1. Diisocyanate Considerations

The structure of the diisocyanate can have a critical influence on the properties of thermoplastic polyurethanes. The study carried out by Schollenberger[17] is particularly informative in this respect. Table 2 clearly shows that high symmetry and rigidity in p-phenylene diisocyanate leads to polymers with a high modulus. Also, the intermolecular forces are of such a strength that the segmented copolymer is usually not soluble in solvents such as dimethyl formamide if the hard segment content is above 30–40 wt%. However, the additional flexibility in MDI due to the methylene linkage is reflected in the mechanical behaviour of the MDI-based polyurethanes. These properties are attributed to the symmetry of MDI which is sufficient to allow partial hard segment crystallization. In contrast, the *meta* structures in *m*-phenylene diisocyanate and 2,4-tolulene diisocyanate (TDI) seriously disrupt the symmetry and the potential for hard segment crystallization and packing, thereby accounting for the lower physical and mechanical properties of these systems — particularly at higher temperatures. In this connection, calorimetric studies of segmented polyurethanes prepared from pure 2,4-TDI and pure 2,6-TDI[18] lead to the conclusion that the symmetrical crystallizable 2,6-TDI promotes regular ordered hard blocks and observable high temperature transitions. On the other hand, the polyurethane derived from 2,4-TDI, which can undergo both head-to-head and head-to-tail placement with respect to the methyl groups, display no transitions above the hard segment T_g.

As another example of hard as well as soft segment variation, MacKnight et al.[19] have carried out DSC studies on the model series of polyurethanes shown below in which x varied from 2 to 10:

TABLE 2

SOME GENERAL EFFECTS OF DIFFERENT DIAMINE STRUCTURES ON THE FINAL PHYSICAL PROPERTIES OF POLYURETHANE ELASTOMERS[a]

Diisocyanate	Structure	Tensile strength (psi)	Ultimate elongation (%)	300% modulus (psi)	T_2 (°C)	DMF soluble
A. p-Phenylene diisocyanate (p-PDI)	OCN—C₆H₄—NCO	3 800	330	3 400	152	No
B. Diphenylmethane-p,p'-diisocyanate (MDI)	OCN—C₆H₄—CH₂—C₆H₄—NCO	5 500	610	1 900	134	Yes
C. m-Phenylene diisocyanate (m-PDI)	OCN—C₆H₄—NCO (meta)	9 000	580	1 400	152	Yes
D. 2,4-Tolylene diisocyanate (TDI)	OCN—C₆H₃(CH₃)—NCO	4 300	680	360	67	Yes

[a] Polytetramethylene adipate glycol was the soft segment utilized in preparing these materials with the chain extender being 1,4 bis(2-hydroxyethoxy)benzine. Taken from Ref. 17.

$$\begin{CD} \end{CD}$$

![Type A structure: -(C(=O)-N(H)-(CH_2)_6-N(H)-C(=O)-O-(CH_2)_x-O-)- Type A]

![Type B structure: -(C(=O)-N(H)-C_6H_4-CH_2-C_6H_4-N(H)-C(=O)-O)_x(-CH_2)_x-O-)- Type B]

![Type C structure: -(C(=O)-N(H)-C_6H_4-N(H)-C(=O)-O-(CH_2)_x-O-)- Type C]

⟨O⟩ = a TDI mixture of the 2,4 and 2,6 isomers

It was shown that the glass transition and crystalline melting points for these high molecular 'homopolymers' was very much a function of the molecular structure. Specifically, Type B displayed the highest T_g values as expected and this is attributed to the rigidity of the aromatic moiety. The T_g values of this polymer were also more sensitive to the number of methylene units in the diol than the all aliphatic Type A derived from hexamethylene diisocyanate (HDI). These values changed from 139°C to 72°C as x changed from 2 to a value of 10. Type B systems also displayed melting transitions which varied with x. In this respect, any relatively minor amount of flexible units introduced into the hard segments will affect the melting, as well as the glass transition, behaviour of the final system.

Although many diisocyanates have been examined, it has been pointed out that few are cheap enough for commercial use and at the same time allow the preparation of polymers that possess adequate physical and mechanical properties.[20] It seems, however, that the highest levels of modulus, tear and tensile strengths are obtained by the use of the most rigid, bulky, symmetrical, and sterically hindered diisocyanates. The diisocyanate structure apparently influences the ability of the hard segments to pack more regularly and consequently to have higher intermolecular hydrogen bond interaction and a stronger physical net-work.[20] It is suggested that the low temperature properties of poly-urethane elastomers are affected only slightly by the type of diisocyanate used.[17] Clearly, however, this statement must at least be based on an equal degree of phase separation since mixing of the hard and soft segments raises the T_g of the soft segment phase and begins to limit low temperature applications.

4.2. Chain-extender Effects

The variation in the chain-extender structure can dramatically influence the elastomeric properties of polyurethane systems. Based on X-ray results, Bonart and co-workers[21,65-67] suggested that polyurethanes composed of specific macroglycols, butanediol and MDI in the molar ratio 1:1:2 had mutually soluble hard and soft phases and showed no evidence of phase separation. However, comparable diamine-extended polymers were phase separated at the same molar composition ratio. Three-dimensional hydrogen bonding, made possible by excess NH groups in the urea linkage, was proposed to explain the observed data. This molecular arrangement enhances the structural stability of urea-based systems and accounts for their higher softening temperature over comparable diol-extended polyurethanes. In general the employment of a diamine results in polymers with better physical properties than when comparable diols are used, owing, in large measure, to the introduction of a urea linkage which enhances hydrogen bond interaction and provides maintenance of properties to a higher temperature.

More recently, the effect of different diamine extenders has been systematically studied.[22] The diamines investigated are listed in Table 3 and their chemical structures shown. In general, the extenders can be divided into one of three groups. Those that are symmetric (p-PDA, 1,5-DAN, 2,7-DAF) provide high modulus and high tensile strength that persist at high temperatures (150°C). The asymmetric diamines (o-PDA, 2,3-DAF, 2,4-DPt, MOCA), on the other hand, tend to display lower moduli as well as lower ultimate strengths. In contrast to asymmetric diamine-extended polymers, symmetric diamines produce materials that display low elongation at lower temperatures but which increase as the temperature is increased. Bulky but asymmetric diamines (3,6-DAA) display intermediate properties. It is also possible that symmetric dia-mines promote hard segment crystallization so that a stronger physical network is obtained. This may help to explain the high modulus, high tear strength, and lower rate of stress relaxation displayed by the

TABLE 3

SOME EXAMPLE CHAIN-EXTENDER DIAMINES THAT HAVE BEEN STUDIED WITH
RESPECT TO THEIR INFLUENCE ON BULK PROPERTIES[a]

Chain-extension agent	Polyurethane nomenclature	Chemical structure
p-Phenylenediamine	p-PDA	
p-Phenylenediamine	p-PDA (10 NCO)	
p-Phenylenediamine	p-PDA (15 NCO)	
o-Phenylenediamine	o-PDA	
1,5-Diaminonaphthalene	1,5-DAN	
1,5-Diaminonaphthalene	1,5-DAN (10 NCO)	
1,5-Diaminonaphthalene	1,5-DAN (15 NCO)	
3,6-Diaminoacridine	3,6-DAA	
3,6-Diaminoacridine	3,6-DAA (10 NCO)	
3,6-Diaminoacridine	3,6-DAA (15 NCO)	
2,7-Diaminofluorene	2,7-DAF	
2,3-Diaminofluorene	2,3-DAF	
2,4-Diamino-6-Phenyl-triazine	2,4,DPT[b]	
3,3'-Dichoro-4,4'-diaminodiphenyl-methane	MOCA[b]	
1,4 Butanediol	1,4-BD[b]	

[a]Modified from Ref. 22.
[b]Catalysed with 1% stannous octoate.

elastomers extended with a symmetric diamine. On the other hand, the asymmetry of the chain extender disrupts hard segment regularity and interferes with hard segment intermolecular association; consequently, lower physical properties are obtained. Bulky diamines, though asymmetric, display intermediate behaviour owing to the high chain stiffness and reduced chain flexibility. In general, aromatic diamines that possess bulky or symmetric structures will give optimum physical properties depending on the material application.

Many diamines are too reactive for use as chain extenders in the preparation of bulk elastomers, as insufficient time is available for adequate mixing and pouring before gelation occurs. However, this limitation is not always a major obstacle for systems that are prepared by reaction injection moulding. Here diamines offer the rather unique combination of desired physical properties, rapid demoulding times, excellent integrity at demoulding and improved release characteristics.[23] The same comments hold in the case of Spandex fibres which are reaction spun from solution.[20] In these systems, aliphatic diamines, such as ethylene diamine, are often chosen as chain extenders since they combine a high-speed continuous process for preparation and a good level of physical properties.[20]

Linear aliphatic diamines offer another variation in the preparation of segmented polyurethanes. It has been claimed that the properties of such systems show a dependence on the odd–even nature of the diamine moiety.[24,25] Specifically, tensile properties are higher when the diamine has an odd number of carbon atoms in the chain,[24] while the tensile modulus and the softening temperature of the hard segment domains are highest for diamines having an even number of carbon atoms.[25] This latter result is attributed to the weaker physical network that results from random packing of hard segments due to steric hindrances of the molecular chain in systems based on diamines with an odd number of carbon atoms. As a result, the former systems display a lower modulus and higher strain-to-break, which in turn allows further stress-induced crystallization of the soft segment (if possible) and consequently higher tensile strength.

More recently, X-ray diffraction and conformational analysis were used to study the bulk structure of MDI/glycol/ester polyurethanes.[68] It was concluded that for butanediol and longer diol extenders, the hard segment structure depended on the even–odd number of methylene groups in the glycol backbone. The even diol polymers adopt the lowest energy, fully extended conformation that allows for hydrogen bonding in both directions perpendicular to the chain axis. Such hydrogen bonding is not possible for the odd diol polymers in the extended conformation and these adopt staggered chain structures having triclinic unit cells with lower crystalline order than that of the even glycol extenders. On the other hand, the shortness of ethylene glycol and 1,2-propylene glycol force such systems to adopt contracted unstaggered structures. The researchers proposed this interpretation as a possible means of obtaining overall better properties of butanediol-based elastomers.

The properties of polyurethanes extended with different linear aliphatic glycols are very much dependent on the number of carbon atoms in the glycol moiety.[4] As indicated from the data in Ref. 4, the 300% modulus as well as the tensile strength shows a maximum as a function of the number of carbon atoms in the extender.[26] Similar conclusions were made by Critchfield et al.,[27] who studied a systematic series of linear aliphatic glycol extenders ranging in methylene content from 2 to 12. The polyol they employed was 2000 M_n polycaprolactone and the diisocyanate was MDI. A constant molar ratio (1:2:1) of polyol:diisocyanate:diol was used unlike the study cited earlier.[68] It was observed that the modulus values show a dependence on the glycol length, with the elastomers containing 1,6-hexanediol and 1,7-heptanediol displaying the lowest moduli. The observed change in mechanical stiffness was attributed to a change in both the degree of phase separation and intermolecular interactions, i.e. a significant point of argument was that as the number of methylene units increases, the degree of mixing of the hard and soft segments also increases owing to the greater compatibility of the hard segment with the soft segment.

Permanent set studies carried out on urethanes with different glycol extenders show that ethylene glycol-extended systems show the lowest hysteresis — possibly an important conclusion for fibre applications.[26] It should be pointed out, however, that the relatively small differences in the effect of these extenders is undoubtedly due to their structural similarities.[20]

Additional studies carried out on systems where rigid and bulky glycols are employed as chain extenders lead to the same conclusions as when symmetric and bulky diamines are utilized as extenders.[28] In general, symmetric, rigid and bulky chain extenders provide polymers with better mechanical properties. However, gross asymmetry disrupts hard segment crystallization or packing, which leads to the formation of a weaker physical network, and poorer physical properties will be obtained.

5. EFFECT OF SOFT AND HARD SEGMENT VARIATION ON THE PROPERTIES OF SEGMENTED POLYETHER/ESTER COPOLYMERS

The effect of soft segment type and molecular weight on the properties of segmented polyether/ester copolymers has been investigated by Wolfe.[34] However, the data are complicated by the fact that the reported inherent viscosities of the samples, and consequently their molecular weights, were a function of the soft segment utilized in the preparation process. Correlations were made between inherent viscosities and mechanical properties but it was pointed out that at any given inherent viscosity, the copolymer based on poly(tetramethylene oxide) (PTMO) showed greater tear strength than polymers where poly(propylene oxide) (PPO) or poly(ethylene oxide) (PEO) were the soft segments. This was attributed to the facilitated strain-induced crystallization of the PTMO units. The differences between copolymers, with regard to synthesis and properties, owing to different polyether structures were more pronounced at high soft segment content or when the terephthalate hard segment was replaced by methylene isophthalate. At high soft segment content, copolymer formation was limited by the same diffusion and degradation process discussed earlier. It seems that the tetramethylene isophthalate hard segment units yield soft elastic polymers of low moduli.

Low temperature properties were in many cases dependent on the polyether structure for a given composition. Brittle points for a concentration of 60% hard segment were as follows: PTMO-based system −70°C, PEO copolymer, −58°C, PPO copolymer, −34°C. It seems that brittle points, as well as low temperature properties, are principally a function of the soft segment T_g and flexibility, as might be expected.

In the same study,[34] mechanical properties were more sensitive to the variation in hard segment concentration. In general, for a given copolymer, tear strength, 100% modulus, and the melting point increased as the hard segment fraction increased. The increase in the T_g values of the amorphous phase in PTMO-based copolymers, as the hard segment fraction increased, was attributed to higher concentrations of uncrystallized hard segment units in the amorphous phase, the increased number of crystalline regions, and the greater reinforcement of the amorphous phase by the longer polyester segments.

The water absorption properties were also studied as a function of the polyether structure.[34] Polar molecules where the oxygen-to-carbon ratio is high (such as PEO) are considerably more hydrophilic than PTMO or PPO copolymers. Consequently, PEO-based systems have higher swelling in water and greater susceptibility to hydrolytic degradation than analogous PTMO- and PPO-based copolymers.

The effect of soft segment chain length was also investigated by Wolfe.[34] At a given composition, the longer the polyether chain length, the greater the number of hard segment polyester units. Consequently, the properties of these copolymers will be very much a function, not only of the soft segment length, but also of its type, and the structure and length of the crystallizable hard blocks. In general, the melting point and the mechanical properties show two distinct behavioural characteristics. The copolymers derived from PPO, PEO, and PTMO show the highest melting points, and the highest tear strengths, but display the greatest tendency to stiffen at low temperatures, a behaviour that was attributed to the considerable decrease in final amorphous polyether content as a result of its crystallization during the final stages of polymerization. Consequently a higher concentration of hard segments was obtained relative to the theoretical composition. On the other hand, the behaviour at the greatest chain lengths was dependent on the soft segment type. Polymers derived from PTMO still showed good properties which was attributed to strain-induced crystallization of the soft segments. Polymers based on PPO displayed poor physical properties when the soft segment molecular weight was increased. This unusual combination of properties was attributed to the high incompatibility of the PPO with the hard segments so that two separate and mutually insoluble phases coexist during the polymerization process. This incompatibility is dependent on the PPO chain length and increases as the molecular weight of the PPO oligomers increase.

The polymers based on PTMO displayed a different behaviour. At the longest PTMO chain lengths, the stiffening temperatures and the moduli of the copolymers increase after passing through a minimum at intermediate chain lengths. This is attributed to the partial crystallization of the PTMO units as the soft segment molecular weight increased. However, the longest chain PTMO-based system, unlike PEO copolymers, show no increase in tear strength and display low tensile properties which is explained in terms of the partial phase separation of the melt during synthesis.

5.1. Effect of Molecular Weight and Polydispersity on the Structure/ Property Relationships of Segmented Systems

It has been suggested that the polymerization procedure has an effect on the properties of segmented copolymers. Peebles[69] has theoretically demonstrated that under ideal and stoichiometric conditions with complete conversion, the sequence length distribution of hard segments in segmented polyurethanes follows the most probable distribution. Also a two-stage polymerization tends to result in a narrower distribution of hard blocks than does a single-stage process of the same stoichiometry. It is suggested that this occurs when there are different reactivities for the isocyanate groups of the diisocyanate difunctional monomer. The existence of different reactivities may arise owing to factors such as steric hindrance, alteration of the induction-residence condition of the molecule when it is partly reacted, or the molecular configuration in the transition state.

The effect of the hard and soft segment size distribution on the properties of segmented elastomers has in fact been studied by Harrell,[70] who prepared and investigated the mechanical behaviour of the following systems:

where n is 1, 2, 3 or 4.

These segmented elastomers contain no NH groups and so no hydrogen bonding of the form common in the segmented polyurethanes is possible. The melting points of the above systems increased with increasing size of hard blocks and asymptotically approached the melting point of the homopolymer with the same composition of the hard block repeat units. However, the heats of fusion of the crystallizable hard block component of the system were found to be approximately constant suggesting that hard block crystallinity was independent of the hard segment length. Various blends of the individual polymers containing monodisperse hard blocks were made by solution-casting pairs of these polymers ($n = 1, 2, 3, 4$), using chloroform as the solvent. Blends of these polymers with hard segments, where n is 1 or 2, gave clear films while turbidity was observed for those with $n = 3$ or 4. Fusion peaks of component polymers were noted by DSC. All the clear films had heat of fusion peaks characteristic of the two individual hard segments present. In addition, at least two new peaks were observed, speculated to be caused by depressed melting points, indicative of the incompatibility of the two hard segments of different sizes. The turbidity observed in the films was later shown to be caused by spherulitic superstructures developed within these systems, which could be controlled by sample preparation.[71] Complementary mechanical tests showed that when the hard segment molecular weight distribution was varied by mixing the components, the materials with a narrow distribution gave polymers of significantly higher modulus than those having a broad distribution. Narrow distribution of hard and soft segments produced polymers with greater tensile strength and greater elongation at break. This behaviour was attributed to a more perfect physical network when either or both of the segments have narrow molecular weight distributions. On the other hand, hysteresis or permanent set was lower when the size distribution for either segment was broad. This was particularly more pronounced for the case of the hard segment distribution. It was also suggested that in these systems strain-induced crystallization of the soft segments may occur, thereby possibly affecting the viscoelastic recovery of the material.[2]

Extensive dynamic mechanical property studies[72] have also been carried out on these same urethanes prepared by Harrell. The data show differences again between narrow and broad hard segment molecular weight distributions and based on these data it has been suggested that the hard segment domains of broad molecular weight were more diffuse since they were diluted with interpenetrating soft segments.

The effect of soft segment polydispersity on the structure of ester-based segmented urethanes has also been considered on the basis of mechanical studies.[73] Conclusions were developed concerning the degree of mixing of hard and soft components but within these same systems the additional variable of significant hard segment crystallinity existed, which is not generally as prominent for the conventional urethanes. It is the authors' belief that there remains a need for more work regarding the effects of soft or hard segment distribution on mechanical and structural behaviour.

The effect of the method of polymerization (one-stage versus two-stage) has also been systematically studied with regard to influence on mechanical properties for at least two systems.[74] The urethanes investigated were 1000 M_w polytetramethylene-adipate and polytetramethylene-oxide diols, while the hard segment was MDI and butane-diol (approximately 20% by weight). Combined X-ray and mechanical analysis suggested that the polymerization method is more significant in the case when the soft segment is a polyester. Specifically the one-stage polyester displayed lower stress, higher hysteresis and a lower degree of phase separation than its two-stage counterpart, while the ether-based system showed little appreciable differences in these same characteristics for the one- and two-stage materials. It was concluded that the observed behaviour was caused by the higher degree of hydrogen bonding and the polyester-based systems brought about by stronger interaction between the urethane NH and the carbonyl of the polyester segments. The broad molecular weight distribution of hard segments resulting from a one-stage polymerization process promotes the solubilization of the shorter segments or isolated diisocyanate moieties within the polyester matrix. This seems to be largely offset by the higher hard–soft segment incompatibility in the case where the soft segment is a polyether. In brief, since MDI is a symmetrical diisocyanate, having the isocyanate groups of rather equal reactivity, the study mentioned above principally demonstrated that the mode of polymerization had little effect on final material properties. This result is in line with the basic theory of Peebles.[69]

Considering processing with respect to final properties and structure, the reaction injection moulding process has been demonstrated to have a profound effect on the structure/property relationships of segmented polyurethanes. Macosko and co-workers[44] have demonstrated that owing to the exothermic nature of the copolymer-forming reaction, a gradient in morphological features may vary across the mole thickness direction. This results from the gradation in reaction temperature caused by limited thermoconductivity of the system. This may also lead to a gradient in molecular weight distribution and segment distribution of the resulting copolymers. In particular, these workers addressed two particular aspects, the first being the heat generation–heat removal balance consideration leading to higher molecular weight polymer at the centre of the mould in contrast to the material found near the mould surface. (A more uniform molecular weight tends to be obtained by preheating the mould or by a post-curing reaction.) The second factor they have addressed is related to the phase separation process which may occur during polymerization; this in turn leads to a broader MWD and may of course influence the morphology and final bulk properties.[44]

It has also been demonstrated that morphological characteristics of RIM systems vary with polymerization temperature and can also vary across the thickness direction of a RIM part as suggested above. Higher polymerization temperatures promote hard segment organization in the as-polymerized material.[44] In brief, it seems, however, that the morphology of these systems is not only a function of the formulation utilized but is also dependent on the flow rate of the liquid components, the reactivity, the catalyst level, the mould thickness and, possibly, the presence of filler particles.[45]

Some further comments relating to past-processing annealing and its effect on properties will be given later. It should be pointed out, however, that most studies of RIM urethanes have been carried out on crystallizable materials. The authors believe that the above discussed conclusions will not necessarily apply to materials that are totally amorphous.

5.2. Effect of Molecular Weight on Structure/Property Behaviour

Schollenberger and Dinbergs[36,37] have investigated the effect of increasing the number and weight average molecular weight of the total polymer on the structure/property relationships of segmented polyurethanes. The hard segment utilized was MDI extended with butanediol (the total being 20% for the M_w study and 38% for the M_n study) and the soft segment was polytetramethylene adipate possessing a M_w of around 1100. The polymer was made by a one-stage reaction process and the M_w varied from 48 000 to 367 000 as determined by GPC and had a broad MWD that increased as the M_w value increased. Characterization and mechanical property measurements on the polymers of this series showed that the change (enhancement or improvement) of the polyurethane properties with increasing M_w continued to a certain M_w level, in the range of 100–200 × 10^3, then tended to level off. The point at which this levelling occurred was referred to as the 'polymer inflection molecular weight' (IM_w). This behaviour was attributed to the achievement of a polymer average chain length which: (1) favours a more random coil (less extended) chain configuration; (2) produces a virtual crosslinked network which is unresponsive to further chain length increase; and (3) has a chain end (free volume) content where further reduction does not affect polymer internal segmental mobility and thus polymer morphology and related density.

Results show that as the polyurethane molecular weight increased, several characteristic properties increased. These were (approximate IM_w): specific gravity (180 000); processing temperature utilizing a dynamic extrusion rheometer (25-mil. samples, 145 000; 75-mil. samples, 120 000); abrasion resistance, Tabor low temperature modulus in the range −70–25°C, Gamen temperature (112 000); and Clashburg temperature (temperature at which the apparent rigidity modulus of the test sample is 45 000 psi) (134 000). It was also noted that as the polyurethane M_w increased, several polymer characteristic properties decreased. These were: mechanical hysteresis (125 000); extension set (130 000); stress relaxation flex life (180 000 with a maximum flex life at 100 000). Polymer hardness in this study varied little with M_w and neither hardness nor the 200% modulus showed a systematic dependence on this same parameter.

5.3. Hydrogen Bonding Effects and Polyurethanes

In contrast to the segmented polyesters, the stronger secondary hydrogen bonding that may exist in the segmented urethane or segmented urea materials can influence the morphological texture and properties of these latter systems. Early studies in the literature strongly imply that hydrogen bonding played a tremendously important role in the structure/ property behaviour of the segmented urethanes. While indeed hydrogen bonding is of significant importance, it is likely that it does not play as strong a role as was initially believed. Below a few of the highlights presented within the more recent literature will be considered.

Data have been presented by Wilkes and co-workers[75,76] and Seymour and co-workers[77-79] on polyurethanes that are not capable of hydrogen bond interaction owing to the fact that the nitrogen is within a piperazine ring. Their results suggest that the usual properties of segmented polyurethanes are principally caused by the inherent incompatability of the polymer components and the occurrence of microphase separation. It seems, however, that hydrogen bonding may well affect the extent of phase separation in related systems and consequently their mechanical properties.[41,75-79] In conventional polyurethanes, hydrogen bonding occurs between secondary NH and carbonyl groups in the hard segments, while hard to soft segment interaction occurs between urethane NH and with either the ether oxygen of polyether soft segments or the carbonyl of the polyester soft segments.[78,80]

Seymour and Cooper[77,79] studied the thermal mobility of polyurethane hydrogen bonds by following the temperature dependence of infra-red absorption for the NH vibration. The hydrogen bond interaction reduces the NH vibrational frequency and increases its intensity. The NH moieties participating in hydrogen bonding show a peak at $3320 \, \text{cm}^{-1}$ whereas non-associated NH absorbs around $3450 \, \text{cm}^{-1}$ with a significantly smaller absorption coefficient.[41,81,82] Infra-red absorption frequencies, as well as intensity measurements, should give a qualitative measure of the degree of hydrogen bonding with segmented urethanes. It has been shown that a slope discontinuity in the absorbance temperature curves occurs at around 80° in several systems and in the past this has been attributed to hydrogen bond disruption at the glass transition temperature of the hard segments.

Senich and MacKnight[83] have investigated the nature of hydrogen bonding and its temperature dependence for a 40% hard segment (2,6-TDI & 1,4-butanediol) and 2060 M_n polytetramethylene oxide as the soft segment using the FTIR technique. Investigation of the NH and carbonyl bond stretching absorptions indicate that 88% of the NH groups and 60% of the carbonyl groups are hydrogen bonded. The temperature dependence of the concentration of hydrogen bonding groups for this study did not show any discontinuity at a temperature considered to be the hard segment T_g. An additional absorption band in the carbonyl stretching region at 1726 cm^{-1} was attributed to an oxidation product of the polyether segment formed during initial sample preparation and subsequent thermal treatment. It should be pointed out, however, that the conclusions made above, regarding the degree of hydrogen bonding, apply solely to the specific formulation of the samples prepared by Senich and MacKnight; this is also possibly true for some of the other speculations made by other specific workers concerning their own samples. In brief, the distribution of hydrogen bonding between carbonyl and NH groups is undoubtedly a function of composition, segment and chain-extender types and the molecular weight of the system, and therefore the nature and degree of hydrogen bonding will likely depend on the specific system studied.

Sung and co-workers[84] have investigated the effect of increasing soft segment molecular weight on mixed segment hydrogen bonding, using infra-red spectroscopy and infra-red dichroism techniques. It was concluded that increasing the soft segment molecular weight from 1000 to 2000 resulted in a higher degree of phase separation and consequently a more perfect domain structure — this conclusion being later supported by SAXS studies of the same system.[85]

Based on wide and small angle X-ray scattering results, Bonart and co-workers[65-67,86] have proposed models of molecular arrangements of the hard segment chains in segmented urethanes and segmented urethane–ureas. Domain structures have been related to the mutual hard segment affinity which allows the packing of hard segments, such that a three-dimensional network of a maximum number of stress-free hydrogen bonds exists. Softening temperatures were correlated with hard segment molecular arrangements and the degree of hydrogen bonding. Diamine-extended polyurethane showed a higher degree of phase separation over comparable urethane formulations extended with butanediol, thereby producing urethane linkages. Three-dimensional hydrogen bonding made possible by excess NH groups in the urea linkage was proposed to account for this difference.[67] More recent studies of hard segment ordering have been carried out by Blackwell and Gardner using low molecular MDI-type analogues.[87] Unfortunately, direct application of these data to polymeric systems has, as yet, not been directly demonstrated, but it is anticipated that these model studies will provide further insight into the hard segment packing characteristics and hydrogen bonding. Finally, a very recent and new domain model derived by SAXS analysis, has been proposed by Koberstein and Stein for some rather typical segmented systems.[88] This model depicts the hard segments as being able to display more flexibility, leading to re-entry or a folding back of this hard segment into the domain. While the model remains to be further investigated, it does promote new thoughts regarding possible domain texture.

5.4. Morphological Investigations of Segmented Polyurethanes and Polyester Segmented Copolymers

Over the last few decades, there has been an interest in attempting to observe the domain or microphase separation morphological texture that arises within the segmented copolymers. In contrast to the long block diene copolymers, such as those of the styrene/butadiene/styrene type, the urethanes and segmented esters tend to be much less susceptible to high electron dense staining by such systems as osmium tetroxide, thereby hindering direct thin film transmission electron microscopy studies of their two-phase morphology. Nevertheless, there have been several attempts to utilize transmission microscopy to investigate these segmented copolymers using different methodologies. For example, Koutsky et al.[89] attempted to study polyester- and polyether-based segmented urethanes by using iodine as a staining agent which it was suggested would be deposited mainly within the hard phase. While claims were made for observance of domain textures, there was some doubt as to whether the iodine was preferentially depositing as speculated. In particular, the domain sizes tended to vary considerably and were larger than what would have been anticipated from other structural studies such as SAXS. Samuels and co-workers[75,76] also utilized transmission microscopy techniques to observe the crystalline domain texture within the earlier piperazine-based urethane materials that Harrell had produced. Again, while the micrographs tend to suggest domain texture, there is some concern as to whether the microscopy method allowed direct observation of the domains and therefore still left some question as to their specific morphological texture. More recently, Fridman, and

Thomas[90] studied the morphology of specially made urethanes where the soft segment was 2000 M_n atactic polypropylene oxide and the hard segment was MDI extented with 1,4-dihydroxybutene-2 (BEDO). The existence of a double bond in the BEDO moiety allowed for selective staining of the hard segment with osmium tetroxide. Electron micrographs suggested that the hard segments formed paracrystalline domains, which were fibrillar in nature, and that the fibres were arranged radially into a spherulitic-like structure. The concentration of the hard segments also tended to be greater at the centre of these superstructures, suggesting possible preferential agglomeration of molecules with longer hard segment sequences at the onset of phase separation as the solvent that was used for solution casting was allowed to evaporate. Other workers have also attempted to apply the TEM methodology but in general at this time it is felt by the present authors that this method has not conclusively brought forth the detailed nature of the domain texture in urethanes for direct observation.

Within the segmented polyesters however, there seems to have been more success in delineating the type of morphological features. For example, Cella and co-workers have used this methodology for investigating the Hytrel copolyesters and their data seem to strongly suggest rather clear observation of the two-phase structure.[32,33] More recent TEM studies by Cooper and co-workers also present detailed morphological information on related systems.[91-93] A difference here, however, is that in contrast to the segmented urethanes, the hard segments of the segmented polyesters tend to be of significant length which promotes possible chain folding and lamellar texture development, thereby providing a more discrete crystalline phase than that typically believed to exist in the segmented urethanes. It may be recalled, however, that Koberstein and Stein recently have speculated that the conventional MDI–BD hard segments may display some folding or larger scale hard segment flexibility than has been suggested up to now.[88]

Superstructure development within either the urethane or ester–ether segmented systems has been well observed through the use of either microscopy methods or small angle light scattering (SALS) methodology. One such example of some spherulitic structure observed in the Harrell series of piperazine-based urethanes is shown in Fig. 3, as noted by the use of SEM. Optical microscopy along with SALS demonstrated that these systems were anisotropic, as would be the polyolefin type of spherulites such as polyethylene. The difference, however, is that these segmented urethanes are generally believed not to give rise to the same

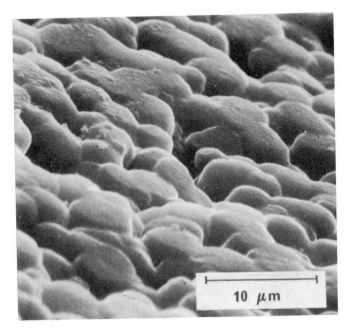

FIG. 3. Scanning electron micrograph of a solution case film of a segmented polyurethane utilizing piperazine as a hard segment. This particular polymer is referred to as N4 in Ref. 71.

nature of chain folding that is possible in the more common spherulitic materials. Chain folding, however, may be possible in the segmented polyester materials owing to the nature of longer and more flexible hard segments. Possible morphologies suggested for some of these spherulitic segmented polyurethanes are shown in Fig. 4, whereas those which tend to allow more chain folding are shown in Fig. 5, the latter having been proposed by Cooper and co-workers.[91-93] Additional references alluding to the superstructural investigations on segmented polymers can be found in Refs 94–98. It is to be stated here, that the use of the SALS technique has also been particularly beneficial for studying the development of the anisotropic superstructure within these materials, as well as its deformation.

5.5. Effects of Crosslinking on Segmented Polyurethanes

Since the authors are not aware of specific studies concerned with crosslinking effects on the segmented polyester materials, emphasis will

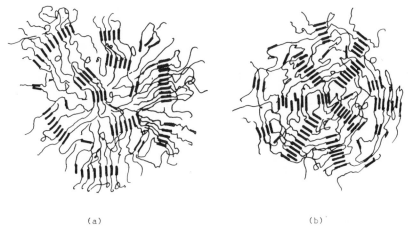

(a)   (b)

FIG. 4. Two models that have been proposed to explain the anisotropic spherulitic structure observed in segmented polyurethanes. These models are only two-dimensional and should be viewed as being representative of a centre 'slice' of a spherulite: (a) chains aligned radially; (b) chains aligned tangentially. Taken from Ref. 71.

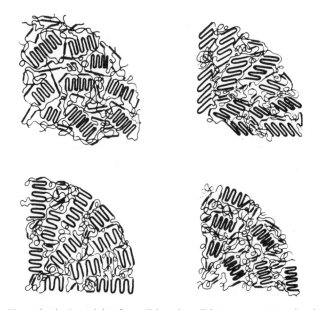

FIG. 5. Hypothetical models of possible spherulitic textures occurring in crystallizable segmented polymers where the hard segment may display considerable chain folding. Darker thicker lines represent the crystalline hard segment. From Lilaonitkul et al.[91]

be placed within this section only on crosslinking effects influencing the segmented urethanes or urethane–ureas. Although the typical urethanes are essentially linear thermoplastics, it is possible to induce some degree of crosslinking via side reactions, like biuret (via urea groups) or allophanate formation (via urethane groups), through the use of organic peroxides, or using soft (or hard) segments and chain extenders that incorporate polyfunctional moieties. Early studies[9,28] regarding crosslinking effects on the mechanical spectrum of polyurethanes showed that upon crosslinking, the modulus in the rubbery plateau region was lowered. This particular behaviour was investigated extensively by Cooper and Tobolsky[99] who suggested that crosslinking within the hard domains prevents optimal packing of the hard segments and reduces the content and strength of the glassy or paracrystalline domains. Similar results and interpretations were reported within another study.[100] Later work[101–103] provided further evidence that crosslinking interferes with domain formation and in certain cases can reduce it greatly.[66] Some studies have also considered the effect of crosslinking on crystallization.[101,104] In brief, it was found that light crosslinking may enhance strain-induced crystallization[101] because the interchain slippage is more limited, while a higher degree of crosslinking can reduce crystallization by imposing restrictions on segmental alignment.

Wilkes and co-workers[105,106] have studied the effect of crosslinking on domain formation for a series of polyester-based urethane elastomers. Organic peroxide was used as a curing agent and the crosslinking reaction was carried out at 200°C — a thermal region where often the hard–soft segment intermixing is high, as discussed in other references.[105–107] Combined X-ray and mechanical studies revealed that the samples with a lower degree of crosslinking displayed better phase separation, as was evidenced by the SAXS analysis and higher Young's moduli. Further studies of these same systems showed that the hysteresis properties of these polymers decreased as the crosslink density increased.[107] In addition, at high elongation, the material with the highest crosslink density displayed a higher stress, the opposite of the behaviour at low strains as illustrated in Fig. 6. It was concluded that at lower elongation the modulus is principally controlled by the degree of phase separation and consequently the linear polymer possesses the highest degree of microphase separation. Upon deformation, however, domain disruption and rearrangement occurs; but covalent bonding is not affected and this accounts for the higher stresses at higher elongations within the crosslinked materials.

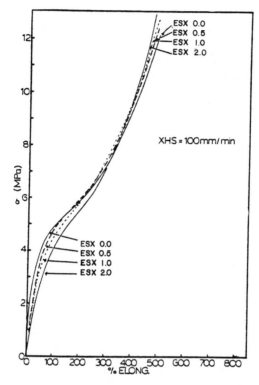

FIG. 6. Plot of stress vs. percentage elongation for a series of segmented polyurethanes based on MDI, butanediol and a polyester soft segment. The numbers refer to levels of peroxide added, i.e. 0·0 indicates no peroxide whereas 2·0 refers to two parts by weight. Note that higher crosslinking promotes a lower modulus at low strains but a higher stress at greater elongations than the uncrosslinked material. Taken from Ref. 107.

Joseph and Wilkes have studied the effect of electron beam crosslinking below and above the melting point of a semicrystalline ester–urethane polymer derived from polycaprolactone.[108] When the prepolymer was cured below the melting point the changes in mechanical properties which occurred as dosage increased are easily attributed to the increase in crosslink density. Isothermal crystallization kinetics measurements showed that the rate of crystallization decreased as the electron beam dosage was raised. Systems that were crosslinked above the melting point exhibit a lower modulus compared with the materials crosslinked in the solid state and this was expected since the former resulted in lower crystallinity and packing order, caused by the network limiting crystallization and superstructure development.

5.6. Effect of Thermal Treatment on the Structure/Property Relationships and Morphological Texture of Segmented Copolymers

Morphological changes in segmented urethane elastomers induced by thermal treatment have been studied by Wilkes and co-workers.[58,105,106,109,110] Additional work has been carried out by Copper and co-workers[111,112] and Jacques.[113] All these workers applied various techniques such as DSC, IR spectroscopy, mechanical analysis, SAXS and WAXS, as well as NMR spectroscopy, to investigate the effects of thermal annealing on morphology.

Cooper and co-workers initially studied the effect of annealing temperature (up to 200°C) and time on the thermal response of polyester and polyether MDI–BD urethanes using DSC analysis. In general, three hard segment associated endotherms were observed in untreated samples at temperatures of 80–100°C, 100–190°C, and above 200°C. These three thermal transitions were correlated with long range order, short range order (which arises in part owing to the distribution of the hard segment lengths) and with the microcrystalline ordering of the hard domains, respectively. Short range ordering could be continuously improved by further annealing; it was suggested that this would produce microcrystallinity development of hard segments only if the hard segment were suitably symmetric and in sufficient content. It was also suggested that the same process could take place without affecting any long range order that might be present. Similar DSC studies by Samuels and Wilkes[75] and Cooper et al.[111,112] on polyurethanes devoid of hydrogen bonding, however, showed virtually the same intermediate transition behaviour and the same behaviour with annealing. These latter studies clearly demonstrated that much of the earlier interpretation of thermal transition behaviour did not arise from changes in the nature of hydrogen bonding as was once suspected.

Jacques[113] also studied the effect of high temperature annealing (150–250°C) on the morphological texture of polyester urethanes. DSC, IR spectroscopy, WAXS and dynamic mechanical property analysis of crystallizable polyurethanes showed that, for his systems, annealing of control samples below 200°C may result in long range ordering of the hard segments but that annealing above 200°C can lead to crystalline hard segment domains, where a melting point was observed at around 249°C with an associated dramatic decrease in Young's modulus. It must be restated here, however, that the effect of thermal treatment on morphological texture will clearly be dependent upon the chemical nature ,of the system, as would be expected, and, therefore, direct utility of some of the above statements as applied to new urethanes of different chemical structure would likely be invalid. The principal point is that thermal treatment can strongly influence morphological texture.

As has been demonstrated by Wilkes and co-workers,[105,106,109,110,114] and in contrast to some of the above statements, many of the segmented urethane systems lose or partially lose domain texture via hard–soft segment mixing when exposed to temperatures above ambient. These workers have investigated the kinetics of phase separation in segmented urethanes following such thermal treatments. Samples were exposed specifically for short periods of time (5–10 min) at temperatures anywhere from 60 to 200°C and then quenched to ambient conditions. These materials, depending on the level of thermal treatment, exhibited significant changes in soft segment T_g, and degree of phase separation (as clearly delineated by SAXS), and an increase in Young's modulus, as a function of the time that the material was allowed to age following quenching. To explain these phenomena it was suggested that the short thermal treatment promotes domain disruption and hard soft segment mixing, as illustrated in Fig. 7. At the annealing temperature the domain morphology is less favoured thermodynamically than at lower temperatures. The soft segments, which also have to be somewhat strained at the lower temperature to allow for phase separation owing to the relatively short length, tend to promote entropically driven stress as induced from rubber elasticity behaviour, as the temperature is raised. This stress will increase as the temperature increases, along with the relative increase in solubility of the two segments — both factors which will favour segment mixing. This tends to promote domain disruption (loss of microphase separation) if a sufficient temperature is reached. Upon cooling, however, the driving force again exists for phase separation, and it is indeed induced but the demixing process is not instantaneous. At room temperature only the soft segments are above their T_g, while the hard segments are glassy and their mobility is necessarily low, thereby hindering the phase separation process. As has been clearly shown by the SAXS method, however, phase separation may occur; this involves the transport processes of the respective segments through the highly viscous matrix in order to promote domain formation. As has been demonstrated by these same workers, covalent crosslinking, molecular weight distribution of the segments, crystallinity

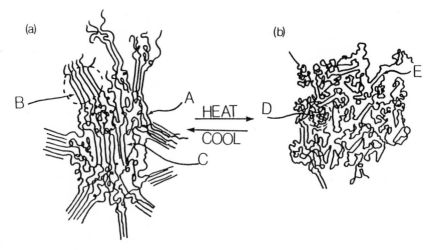

F<small>IG</small>. 7.   Schematic model depicting the morphology at both (a) long time aging and (b) following heat treatment. A, partially extended soft segment; B, hard segment domain; C, hard segment; D, cooled or 'relaxed' soft segment; E, lower order of hard segment domain. Taken from Ref. 110.

of either of the phases, and the temperature following thermal quenching may alter the driving force for phase separation. As would be expected, if the polymer is initially highly ordered, i.e. semicrystalline domains exist, domain disruption at higher temperatures is significantly reduced or may even be absent.[58,112,114] Also, in the case of amorphous urethanes with low hard segment content, the shorter average length of the hard segments renders them even more soluble within the soft segment matrix and this in itself reduces phase separation as might be expected.

Hesketh and Cooper[111] carried out similar time-dependent studies on a number of segmented copolymers at various temperatures (120, 150, 170 and 190°C). All polymers contained 50% hard segment by weight and the same polyether soft segment but different hard segment types. Annealing for 4 h, followed by subsequent quenching, resulted in an exponential decay to the steady state T_g values of the segments. The T_g displacement from its well aged value was greater for higher annealing temperatures as had been observed earlier by Wilkes and co-workers.

Anealing effects have also been noted for semicrystalline segmented ether/ester copolymers.[91] In general, long annealing time (several hours at 150°C) will have some effect on the superstructures produced within these systems. Specifically, there is some control of the spherulitic texture that these materials possess which can be somewhat altered by thermal annealing. Additional DSC studies in these segmented ether/ester systems reveal multiple DSC events which have been correlated to thermal treatment and sample preparation.[35] It might also be pointed out that in cases where the soft segment can crystallize, there may also be time-dependent hardening upon processing owing to the partial crystallization of this phase. In general, however, most of the time-dependent properties often arise from the hard segment domain formation.

6. SUMMARY

In this chapter some of the basic differences in chain chemistry and structure that will promote effects on the final bulk properties of the segmented copolymers that result have been emphasized. Studies addressing the overall structure/property behaviour of segmented copolymers that have been presented within the literature are not exhaustive by any means. Rather, an introduction to this area has been provided, along with the reservation that the reader must be cautious in directly applying the results found from one study to those of their own, where differences in chemistry or processing conditions might exist. However, it is clear that the stage is set with regard to the importance of the segmented copolymers in the industrial sector. They have provided an immense range of materials that will likely grow in scope for man's use and benefit in years to come.

7. ACKNOWLEDGEMENT

The authors would like to express their appreciation for the financial support from the Polymer Materials Division of the National Science Foundation during the preparation of this text.

REFERENCES

1. Treloar, L. R. G. (1974). *Rubber Chem. Technol.*, **47**(3), 625.
2. Noshay, A. and McGrath, J. E. (Eds), (1973). *Block Copolymers*, Wiley, New York.
3. Lagasse, R. R. (1977). *J. Appl. Polym. Sci.*, **21**, 2489.

4. Brunette, C. M., Hsu, S. L., Rossman, M., MacKnight, W. J. and Schneider, N. S. (1981). *Polym. Eng. Sci.,* **21**(11), 668; **21**(3), 163.
5. Ono, K., Shimadu, H., Nishimuro, T., Yamashita, S., Okamoto, H. and Minoura, Y. (1977). *J. Appl. Polym. Sci.,* **21**, 3223.
6. Schneider, N. S. and Matton, R. W. (1977). *Polym. Eng. Sci.,* **19**(15), 1122.
7. Dole-Robbe, J. P. (1937). *Bull. Soc. Chim. Fr.,* **3**, 1978.
8. Graham, S. W. and Hercules, D. M. (1981). *J. Bio. Med. Mat. Res.,* **15**, 349.
9. Saunders, J. H. and Frisch, K. C. (1962). *Polyurethanes, Chemistry and Technology,* Part I, Interscience, New York.
10. Lee, L. J. (1981). *Rubber Chem. Technol.,* **53**, 542.
11a. Meyer, D. A. (1964). In: *Vulcanization of Elastomers,* (Ed. G. Alliger and I. J. Sjothun), Reinhold, New York.
11b. Gruber, E. E. and Keplinger, O. C. (1959). *Ind. Eng. Chem.,* **51**(2), 151.
12. Cluff, E. F. and Gladding, E. K. (1960). *J. Appl. Polym. Sci.,* **3**, 290.
13. Pattison, D. B. assignor to E. I. du Pont de Nemours and Co., US Patent 2 808 391, October 1, 1957.
14. Wright, P. and Cumming, A. P. C. (1969). *Solid Polyurethane Elastomers,* Gordon and Breach, New York.
15. Urs, S. V. (1962). *Ind. Eng. Chem. Prod. Res. Dev.,* **1**(3), 199.
16. Schollenberger, C. S. and Dinbergs, K. (1978). *J. Polym. Sci. Polym. Symp.,* **64**, 351.
17. Schollenburger, C. S. (1969). In: *Polyurethane Technology,* Ch. X, (Ed. P. F. Bruins) Interscience, New York.
18. Sung, C. S. P., Schneider, N. S., Matton, R. W. and Illinger, J. L. (1974). *Poly. Prepr., Am. Chem. Soc., Polym. Chem. Div.,* **15**(1), 620.
19. MacKnight, W. J., Yang, M. and Kajiyama, T. (1968). *Anal. Calorimetry,* Proc. Am. Chem. Soc. Symp., 99.
20. Allport, D. C. and Mohajer, A. A. (1973). In: *Block Copolymers,* (Ed. D. C. Allport and W. H. James), Applied Science Publishers Ltd, London, Ch. 8C.
21. Bonart, R., Morbitzer, L. and Muller, H. (1968). *J. Macromol. Sci., Phys,* **3**(2), 337.
22. Hepburn, C. and Reynolds, R. J. W. (Eds), (1980). *Elastomers: Criteria for Engineering Design,* Applied Science Publishers Ltd, London, p. 323.
23. Ludwico *et al. The Bayflex 110 Series,* Society of Automotive Engineers Car Meeting, Detroit, September 26.
24. Heikens, D., Meijers, A. and von Reth, P. H. (1968). *Polymer,* **9**.
25. Bonart, R., Morbitzer, L. and Rinke, H. (1970). *Kolloid-Z,* **240**, 807.
26. B. F. Goodrich Co., British Patent 1 025 970 (15.7.63); Equip. Can. 739 034.
27. Critchfield, E. E., Koleske, J. V., Magnus, G. and Dodd, J. L. (1972). *J. Elastoplast,* **4**, 22.
28. Pigott, K. A., Frye, B. F., Allen, K. R., Steingiser, S., Darr, W. C., Saunders, J. H. and Hardy, E. E. (1960). *J. Chem. Eng. Data,* **5**, 391.
29. Hoeschele, G. K. and Witsiepe, W. K. (1973). *Angew. Makromol. Chem.,* **29/30**, 267.
30. Hoechele, G. K. (1974). *Chimia,* **28**, 544.
31. Hoeschele, G. K. (1977). *Angew. Makromol. Chem.,* **58/59**, 299.
32. Cella, R. J. (1977). In: *Encyclopedia of Polymer Science and Technology,* (Ed. H. F. Mark *et al.*) Suppl. Vol. II, J. Wiley & Sons, New York, p. 485.

33. Cella, R. J. (1973). *J. Polym. Sci., Polym. Symp.*, **42**, 727.
34. Wolfe, J. R. (1974). *Rubber Chem. Technol.*, **50** (4), 688.
35. Wegner, G., Fujii, T., Meyer, W., and Lieser, G. (1978). *Angew. Makromol. Chem.*, **74**(1204), 295.
36. Schollenberger, C. S., and Dinbergs, K. J. (1979). *Elastoplastics*, **11**, 58.
37. Schollenberger, C. S. and Dinbergs, K. J. (1973). *Elastoplastics*, **5**, 222.
38. Schneider, N. S., Dusablon, L. V., Snell, E. W. and Prosser, R. A. (1969). *J. Macromol. Sci.–Phys.*, **3**(4), 623.
39. Allport, D. C. and Mohajer, A. A. (1973). In: *Block Copolymers*, (Ed. D. C. Allport and W. H. James), Applied Science Publishers Ltd, Londong, Ch. 5C.
40. Trappe, G. (1968). In: *Advances in Polyurethane Technology*, (Ed. J. M. Buirt), John Wiley and Sons, New York.
41. West, J. C. and Cooper, S. L. (1978). In: *Science and Technology of Rubber*, (Ed. F. R. Eirich), Academic Press Inc., New York, Ch. 13.
42. Fridman, I. D., Thomas, E. L., Lee, L. J. and Macosko, C. W. (1980). *Polymer*, **21**, 393.
43. Castro, J. M., Macosko, C. W., Critchfield, F. W., Steinle, E. C. and Tackett, L. P. (1980). *Elastoplastics*, **12**, 3.
44. Tirrell, M., Lee, L. J. and Macosko, C. W. (1979). *Polymerization Reactors and Processes*, (Ed. J. N. Henderson and T. C. Bouton), ACS. Symp. Series, No. 104, p. 149.
45. Hicks, J., Mohan, A. and Ray, W. H., *Can. J. Chem. Eng.*, **47**, 590.
46. Lee, L. J. and Macosko, C. W. (1978). *Soc. Plast. Eng., ANTEC Papers*, **24**, 155.
47. Case, L. C. (1958). *J. Polym. Sci.*, **29**, 455.
48. Macosko, C. W. and Miller, D. R. (1976). *Macromol.*, **9**, 199.
49. Lee, L. J., Ottino, J. M. and Macosko, C. W. (1979). *Soc. Plast. Eng.*, **25**, 439.
50. Becker, W. E. (Ed.), (1979). *Reaction Injection Molding*, Van Nostrand Reinhold Company, New York.
51. Dow Chemical Company, US Patent 4 269 945.
52. Bonart, R. and Muller, E. H. (1974). *J. Macromol. Sci. Phys.*, **10**, 177.
53. Bogarchuk, Y. M., Rappaport, L. Y., Nikitin, V. N. and Apukthina, N. P. (1965). *Polym. Sci., USSR*, **7**, 859.
54. Ophir, Z. and Wilkes, G. L. (1980). *J. Polym. Sci., Phys.*, **18**, 1469.
55. Weisfeld, L. B., Little, J. R. and Walstenholme, W. E. (1962). *J. Polym. Sci.*, **56**, 455.
56. Morbitzer, L. and Hespe, H. (1972). *J. Appl. Polym. Sci.*, **16**, 2697.
57. Apukthina, N. P., Zimine, M. G., Novoselok, F. B. and Myulla, B. E. (1972). *Tr. Meshdunar, Konf. Kauch. Rezine*, **77**.
58. Morbitzer, L. and Bonart, R. (1969). *Kolloid-Z*, **232**, 764.
59. Eastman Kodak Co. British Patent 1 118 731 (29.6.64).
60. B. F. Goodrich Co. US Patent 2 871 218 (1.12.55).
61. Bayer, O., Muller, E., Peterson, S., Piepenbrink, H. F. and Windemath, E. (1950). *Rubber Chem. Tech.*, **23**, 812; (1950). *Agnew. Chem.*, **62**, 157.
62. Union Carbide, Belgian Patent 649, 619 (24.6.63, 28.5.64).
63. Yilgor, I., Shah'aban, A., Steckle, W. P., Jr. Tyagi, D., Wilkes, G. L. and

McGrath, J. E. Segmented organo siloxane copolymers, I. Synthesis of siloxane – urea copolymers, *Polymer* (in press).
64. Tyagi, D., Yilgor, I., McGrath, J. E. and Wilkes, G. L. Segmented organo siloxane copolymers, II. Thermal and mechanical properties of siloxane – urea copolymers, *Polymer* (in press).
65. Bonart, R. (1968). *J. Macromol. Sci., Phys.*, **2**(1), 115.
66. Bonart, R., Morbitzer, L. and Hentze, G. (1969). *J. Macromol. Sci., Phys.*, **3**(2), 337.
67. Bonart, R., Morbitzer, L. and Muller, E. H. (1974). *J. Macromol. Sci., Phys.*, **9**, 447.
68. Blackwell, J., Nagarajan, M. R. and Hoitiuk, T. B. (1982). *Polymer*, **23**, 950.
69. Peebles, L. H. (1976). *Macromolecules*, **9**, 58.
70. Harrell, L. L. (1969). *Macromolecules*, **2**(6), 607.
71. Samuel, S. L. and Wilkes, G. L. (1971). *J. Polym. Sci.*, **11**(5), 369.
72. Ng, H. N., Allegrezza, A. E., Seymour, R. W. and Cooper, S. L. (1973). *Polymer*, **14**, 255.
73. Gianflauio, L., Sumida, Y. and Vogl, D. (1980). *Angew. Makromol. Chem.*, **87**, 1.
74. Abouzahr, S. and Wilkes, G. L. (1984). *J. Appl. Polym. Sci.*, **29**(9), 2695.
75. Samuels, S. L. and Wilkes, G. L. (1978). *J. Polym. Sci., Polym Symp.*, **43**, 149.
76. Wilkes, G. L., Samuels, S. L. and Crystal, R. (1974). *J. Macromol. Sci., Phys.*, **10**, 203.
77. Seymour, R. W. and Cooper, S. L. (1973). *Macromolecules*, **6**, 48.
78. Estes, G. M., Seymour, R. W. and Cooper, S. L. (1971). *Macromolecules*, **4**, 452.
79. Cooper, S. L. and Seymour, R. W. (1973). In: *Block and Graft Copolymers*, (Ed. J. J. Burke and V. Weiss), Syracuse Univ. Press, New York.
80. Ishihara, H., Kimura, I., Saito, K. and Ono, H. (1974). *J. Macromol. Sci., Phys.*, **B10**, 591.
81. Pimentel, G. C. and McClellan, A. L. (1960). *The Hydrogen Bond*, Freeman, San Francisco.
82. Hannon, M. J. and Koenig, J. L. (1969). *J. Polym. Sci., Part A2*, **7**, 1005.
83. Senich, G. A. and MacKnight, W. J. (1980). *Macromolecules*, **13**, 106.
84. Sung, C. S. P. (1980). In: *Polymer Alloys — II*, (Ed. D. Klempner and K. C. Frisch), Plenum Publ. Co., New York.
85. Wilkes, G. L. and Abouzahr, S. (1981). *Macromolecules*, **14**, 456.
86. Bonart, R. and Muller, E. H. (1974). *J. Macromol. Sci., Phys.*, **10**, 177.
87. Blackwell, J. and Gardner, K. H. (1979). *Polymer*, **20**, 13.
88. Koberstein, J. and Stein, R. S. Private communication.
89. Koutsky, J. A., Hien, N. V. and Cooper, S. L. (1970). *J. Polym. Sci., Part B*, **8**(5), 353.
90. Fridman, I. D. and Thomas, E. L. (1980). *Polymer*, **21**, 388.
91. Lilaonitkul, A., West, J. C. and Cooper, S. L. (1976). *J. Macromol. Sci. Phys.*, **B12**(4), 563.
92. Shen, M., Mehra, U., Niinomi, M., Koberstein, J. T. and Cooper, S. L. (1974). *J. Appl. Phys.*, **45**(10), 4182.
93. West, J. C., Lilaonitkul, A., Cooper, S. L., Mehra, U. and Shen, M. (1974). *Polym. Prep. Am. Chem. Soc.*, **15**(2), 191.

94. Chang, P. and Wilkes, G. L. (1975). *J. Polym. Sci.*, **13**, 455.
95. Schneider, N. S., Desper, C. R., Illinger, J. L., King, A. O. and Barr, D. (1975). *J. Macromol. Sci.*, **B11**, 527.
96. Kimura, I., Ishihara, H. and Ono, H. (1971). *IUPAC Macromol. Prepr.*, **23**(1), 525.
97. Kimura, I., Ishihara, H., Ono, H., Yoshihara, N., Nomura, S. and Kawai, H. (1973). *Macromolecules*, **7**, 355.
98. Mody, P. and Wilkes, G. L. (1981). *J. Appl. Polym. Sci.*, **26**, 2853.
99. Cooper, S. L. and Tobolsky, A. V. (1967). *J. Appl. Polym. Sci.*, **11**, 1361.
100. Holden, G., Bishop, E. T. and Legge, N. R. (1969). *J. Polym. Sci.*, **C26**, 37.
101. Kimball, M. E. and Fielding-Russel, G. S. (1977). *Polymer*, **18**, 1777.
102. Nishi, T. (1973). *J. Appl. Polym. Sci., Appl. Polym. Symp.*, **20**, 353.
103. Fulcher, K. U. and Corbett, G. E. (1975). *Br. Polym. J.*, **7**, 225.
104. Miller, G. W. (1971). *J. Appl. Polym. Sci.*, **15**, 39.
105. Ophir, Z. H. and Wilkes, G. L. (1978). *Adv. Chem. Series*, **176**, 53.
106. Assink, R. A. and Wilkes, G. L. (1977). *Polym. Eng. Sci.*, **17**, 606.
107. Abouzahr, S. and Wilkes, G. L. (1980). *Polym. Prep. Amer. Chem. Soc.*, **21**(2), 193.
108. Joseph, E., Wilkes, G. L. and Park, K. (1981). *J. Appl. Polym. Sci.*, **26**, 3355.
109. Bagrodia, S., Wilkes, G. L., Humphries, W. and Wildnauer, R. (1975). *J. Polym. Lett.*, **13**, 321.
110. Wildnauer, R. and Wilkes, G. L. (1975). *J. Appl. Phys.*, **46**, 4148.
111. Hesketh, T. R. and Cooper, S. L. (1977). *Org. Coatings Plast. Chem., Am. Chem. Soc.*, **37**, 509.
112. Hesketh, T. R., VanBogart, J. W. C. and Cooper, S. L. (1980). *Polym. Eng. Sci.*, **20**(3), 190.
113. Jacques, C. H. M. (1977). In: *Polymer Alloys, Blends, Blocks, Grafts, and Interpenetrating Networks*, (Ed. D. Klempner and K. C. Frisch), Plenum Press, New York.
114. Abouzahr, S., Wilkes, G. L. and Ophir, Z. (1982). *Polymer*, **23**, 1079.

Index

L1347 (083)